Sensing Machines

Sensing Machines

How Sensors Shape Our Everyday Life

Chris Salter

The MIT Press
Cambridge, Massachusetts
London, England

© 2022 Massachusetts Institute of Technology

All rights reserved. No part of this book may be reproduced in any form by any electronic or mechanical means (including photocopying, recording, or information storage and retrieval) without permission in writing from the publisher.

The MIT Press would like to thank the anonymous peer reviewers who provided comments on drafts of this book. The generous work of academic experts is essential for establishing the authority and quality of our publications. We acknowledge with gratitude the contributions of these otherwise uncredited readers.

This book was set in Stone Serif and Stone Sans by Jen Jackowitz. Printed and bound in the United States of America.

Library of Congress Cataloging-in-Publication Data

Names: Salter, Chris, 1967- author.
Title: Sensing machines : how sensors shape our everyday life / Chris Salter.
Description: Cambridge, Massachusetts : The MIT Press, [2022] | Includes bibliographical references and index. | Summary: "Sensing Machines shows how sensors and artificial intelligence transform our lives, from driving and playing games, to the ways we eat, sleep and dream"— Provided by publisher.
Identifiers: LCCN 2021011597 | ISBN 9780262046602
Subjects: LCSH: Detectors—Popular works. | Electronic apparatus and appliances—Social aspects—Popular works.
Classification: LCC TK7872.D48 S25 2022 | DDC 681/.2—dc23
LC record available at https://lccn.loc.gov/2021011597ISBN: 978-0-262-04660-2

10 9 8 7 6 5 4 3 2 1

Contents

Prologue 1

1 Measuring Sensation 15

I Playing

2 Accelerate! 43
3 "You Are the Controller" 61

II Immersing

4 Borderless 83
5 Androids That Sing 107

III Engineering

6 Sensing on Wheels 127
7 Tasting Machines 147

IV Monitoring

8 Ten Thousand Steps 165
9 Machines Like Us 179
10 Becoming Aware 195

V Enhancing

11 Sensing and Hacking the (Soft) Self 219

Epilogue 247

Acknowledgments 257
Notes 259
Index 303

Prologue

The alarm rings at 6:00 a.m. sharp. The clock's built-in atomizer perfumes the air with scent and makes wireless contact with the wrist tracker you wore during sleep. Based on how much you moved and snored during your last six hours of hibernation, a playlist is quickly assembled. While you shower, the blasting strains of Madonna, Rihanna, and Taylor Swift are heard. As you enter your kitchen, a smart speaker announces the temperature and asks if it should turn on the espresso maker, turn off the hall light, and adjust the toaster settings.

You look at your phone's newsfeed. The state of the world on the other side of the globe while you were dormant flashes by. Singapore is requiring its 5.7 million citizens to download a contact-tracing smartphone app to monitor crowd proximity. China is deploying delivery drones armed with thermal sensors, microphones and speakers, ultrasonic emitters, and disinfectant atomizers to enforce the world's biggest quarantine against a newly emerging supervirus. Over the Indian Ocean, radar sensors are strapped onto 169 albatross to detect illegally operating fishing boats.

You scroll. Accusations that your new furniture is sending names, IP addresses, and persistent identifiers to third-party marketers. Plans from a Silicon Valley corporation to turn abandoned US suburbs into smart cities called CleverZones, which have been derailed by community activists who claim the new urban utopia is a gigantic data collection machine. A news item about how pulse oximeters that measure oxygen in the blood exhibit racial bias since they have problems projecting light through darker skin. A Reddit post that a Japanese game company's wearable hoodie that tracks temperature and brainwaves to detect a player's mood has been hacked, with data from millions of teenagers' brainwaves stolen.

Your self-driving car provides that ever-more-important extra hour of work preparation during the endless commute. The front door opens upon approach. The custom-installed AI system scans your pupils and adjusts the climate control and the loudness of the financial news blaring from the digital sound system based on your nervousness.

As the car pilots through the streets and onto the overcrowded expressway, your smart watch buzzes and beats on your wrist, one email flooding in after another. A sea of scrolling numbers and graphs animates the watch's face: heart rate; respiration; sweat levels; NASDAQ rises and falls; DMs from friends; temperature, air pressure, and moisture readings. At one point, lost in a daydream, you briefly gawk at a billboard in the near distance advertising web-based security services. Moments later, ads for security services pop up in your browser.

At work you glare into the screen for the next hours, only interrupted by lunch DoorDashed to your desk. The day passes. The colored mood lighting in the office seamlessly shifts from warm to cold based on how long you occupy your chair and the sequences you type. Every click of the keyboard and movement of the mouse is captured and fed back to you in the form of vibrations on the skin from the smart watch when you miss your benchmark. Online consumption breaks the monotony of contracts, emails, spreadsheets. You download an app that tracks how long it takes you to make decisions in the supermarket, purchase a new fitness tracker, update a social media feed to enable others to locate you. Briefly logging into a banking website during a break results in a strange glitch: the cursor suddenly vanishes during the session.

Some eight hours later, after the routine flurry of calls and online meetings with far-flung project teams in distant locales, a rhythmic pattern pulsed from your watch comes through from a friend you'll soon meet. Before rendezvousing at the gallery, you go for a run to discharge your daily stress. The exercise is punctuated by your smart watch's continual readout of your steps, breathing, and heart rate. As you change your running speed, the biometric sensors built into your earbuds adjust the rhythm and tempo of the electronic beats being piped into your ears.

After showering away your sweat, upon arrival at the newest immersive experience space (another one) with your friend, the gallery attendant places a clear wristband on the arm not occupied by your smart watch and smart disease-tracking armband. You meander through endless labyrinths of light

Prologue

and sound. In one place, a sign asks you to enter a narrow passage and stand in the center. As you follow the instructions, life-sized projections of animated creatures, floating characters, and abstract bursts of light swim on the walls. Colossal 24K, ten-billion-pixel projections of anemone-like flowers follow your movements as changing scents are diffused from the walls and the floor.

Hunger takes over. The new experience restaurant is waiting past the gallery exit. The gastronomic "journey" is designed to last three hours, and the attendant scans you and your friend to gather data to customize the meal. You are paraded through a series of different rooms, one for each course. Cocktails arrive that steam and smell. Tastes extracted from now-extinct flora and fauna are served up as foams, sprays, and spheres. The main courses appear in a space plunged into total darkness, with each dish accompanied by tactile and sonic signatures. Dessert arrives in another space hung with mammoth plastic strawberries and watermelon slices and flashing videos on a dozen wall-mounted screens. Liquid nitrogen–treated **ice cream.** Popsicles wrapped in gels. A one-meter-tall Manchego cheesecake pumped with air closes out the gastronomic overload.

Dancing is finally on the agenda, in the accompanying club housed in the bowels of the gallery. Your wristband allows entrance. You dance in the sweating crowd; the floor beneath you palpitates, changing color and rhythm with your moves. Moving lights in the ceiling zero in on different groups of dancers. Their tracked wristbands light up in dozens of colors that pulse and glow in sync with the DJ's rhythms. The air is tinged with haze and smoke, the afterimages of strobe lights and the pounding beat. Ecstasy after the long lockdowns.

Arriving home two hours later, you wind down. Donning the latest headset that senses and records brainwaves, you glance at the readout on your phone—digitally scrawled waveforms that ceaselessly fall up and down on the screen. Closing your eyes, you alter the signal by falling into a state of "alert calmness." Enlightenment nears. More minutes pass. The readout's sudden jerks and wild oscillations gradually become a smooth, undulating curve. You remove the headset, brush your teeth, set parameters on your wrist sleep tracker, and fall once again into deep slumber.[1]

Welcome to sensory life in the twenty-first century. This story might sound like science fiction. But for an ever-increasing percentage of the planet's

Figure 0.1
Sensory life in the twenty-first century.

socially-economically advantaged populations, it is or will soon be daily reality.

The events in the preceding narrative have something in common. They all involve what I call *sensing machines*—electronic sensors, machine "intelligence," human labor, and material infrastructures interwoven with our living, breathing, and moving bodies that sense, interpret, and act on the world.

This book tells the story of this world, one where more electronic sensors and computers than people now occupy the planet—some thirty to fifty billion in 2020, and expected to be over a trillion soon thereafter.[2] It explores how we interact with these sensing machines through our long-evolved human senses and how these machines "sense" and act on us.

Sensing machines run the planet. From the intimacy of our homes to the outer reaches of the earth, they shift our understanding of space and time, bodies and machines, self and environment. Sensing machines tell us the espresso maker's water is the perfect temperature, guide the Roomba vacuum across a dog-hair infested carpet, trigger airbags in the split-second moment of a car crash, change the orientation of an image when rotating a phone, or adjust energy usage to the time of day or the season at hand. They alter our knowledge of the climate and the planet itself, operating across spatiotemporal sites and scales, from the bottom of the ocean and the depths of Earth's crust to the heights of the thermosphere.

Sensing machines assist in the dramatic transformation of our health, social life, and well-being. They effortlessly reshape and reinvent notions of privacy, personal space, intimacy, and selfhood. Through electronics and mathematics, sensing machines give us new insights about what is happening inside of, outside of, and around our bodies and selves while hiding how this data is captured and used. From architecture and the arts to construction, surgery, security, and travel, their pervasiveness leaves no discipline or practice untouched. In short, sensing machines radically

Prologue

reconfigure what it means to work, research, build, eat, exercise, socialize, heal, procreate, sleep, and dream in the twenty-first century.

⌒

Just reflect for a moment on how *sensors*, basic electronic devices that detect changes in the environment and convert or *transduce* that information into computer readable data, are dramatically reimagining life. In an effort to ward off the COVID-19 pandemic that shut down the world in 2020–2021, governments resorted to contact tracing, using smartphones, electronic wristbands, and the branch of artificial intelligence (AI) called *machine learning* to detect the distances and locations of individuals in order to slow viral spread and thus potentially save millions of lives while also reimagining privacy and personal data.[3]

In the ongoing fight against climate emergency, globally networked biogeochemical sensor arrays are being scattered in the oceans, monitoring whether nations are reducing CO_2 by measuring carbon flux across seasons, under the ice and in surrounding waters.[4] Closer to our bodies, embedded devices in the skin monitor the fluctuation of insulin for diabetics, while ingestible wireless piezoelectric sensors scan stomachs for signs of disease,[5] bringing us a step closer to the cyborg dream of integrating with machines.[6]

Sensing machines reimagine how infrastructures that run cities and economies are conceived, built, and controlled. The Hong Kong-Zhuhai-Macao Bridge, opened in 2018 as the longest private roadway in the world, is breathtaking in the number of sensors it deploys: tilt sensors, high-precision gyroscopes, lasers, displacement sensors, hydraulic load cells, wind speed, temperature, humidity, pressure, air pressure difference, and carbon dioxide detectors.[7]

Set within a strip mall–dotted suburban New Jersey landscape near New York City, the nondescript Equinix NY4 Data Center hosts forty-nine of the world's key stock trading exchanges. Guarded like Fort Knox, with five levels of access security, NY4 utilizes thousands of distributed sensors to monitor temperature, power consumption, and airflow for rows of interlinked servers processing more than a trillion electronic transactions daily.[8] In the building information modeling (BIM) world, systems such as Arup Neuron, a vast AI-driven analytics platform for sensor-augmented building monitoring developed by the global architectural engineering firm Arup, are being compared to the human body. A building's heating, ventilation, and air conditioning (HVAC) system is seen as the lungs, the blood vessels, the pipes, the bones and skin; the architecture and the Neuron software platform are the brain.[9]

While sensing machines enhance our security and privacy, they also gnaw away at it. Doorbells, thermostats, appliances, and loudspeakers double as surveillance devices. Workers' brainwaves are monitored in offices and on factory floors to detect stress and increase efficiency.[10] New technologies in behavioral biometrics, designed to reduce cybersecurity threats, are able to measure hundreds of parameters based on the common sensors in smartphones and computers: the angle at which we hold the devices; patterns of presses on the surfaces, the speed of rotation (to change things from landscape to portrait) and mouse movement, and how fast applications are opened and closed.[11]

At the same time, supposedly "objective" sensors perpetuate long-engrained human biases. Soap dispensers using infrared (IR) sensing to detect motion under a faucet favor lighter skin. Facial recognition software mainly identifies white faces. Alexa and Google Home's microphones discriminate against Indian and Black English.[12]

With sensors, even our emotional and mental states are up for grabs. Strap on a new wearable "mood tracker" and find out whether you are feeling happy or grumpy based on how much you sweat. Capture something about your mental state with a "brain-computer interface" that senses your brainwaves. Or, be prepared for a future surprise when you book a flight. Seats wired with heart-rate monitors are currently being prototyped that will sense and display a passenger's mood—a useful feature to head off potential altercations with flight attendants if you become an unruly customer.[13]

But while billions of sensors and algorithms are monitoring and calculating our world in opaque and baffling ways, something else is also occurring. New and dizzying relationships are being formed among our bodies, sensing machines, and the environments we inhabit, in which stimuli have been precisely engineered through these very machines in order for us to reach new sensatory highs: atomizers spray scents in our homes, offices, and lobbies; sleep machines with light sensors and microphones adjust their output according to our sleep patterns; flicker-based LED relaxation masks reduce our stress; brainwave interfaces that trigger light and sound seek to reveal inner brain states; cars with cameras scan our eye movement and monitor our nervousness; immersive artworks track and monitor us to create heightened states of audiovisual vertigo and bliss. Even food, which is now put through a scientific regimen of taste modulators, dehydrators, and liquid nitrogen baths under the term *gastronomic engineering*, is being transformed. In other words, ever-new feedback loops emerge between us and technologies that produce new sensations while monitoring and measuring them.

Sensor-based surveillance seems at first to have little to do with brainwave interfaces for personal transformation or food converted into laboratory experiments. Yet in our contemporary "experience economies," these things are not opposed. The pleasure of sensation and the ruthlessness of impersonal measurement go hand and hand. To put it bluntly, in contemporary life there is no economy without the senses. In partnership with these machines, our human senses are being jacked up; they are on steroids.[14]

As these examples show, sensing machines have become all-powerful in contemporary life. They devour vast amounts of the world's data while deploying mathematical concepts and models that most people had never heard about until recently, like artificial intelligence, machine learning, and predictive analytics that process and predict at lightning speed, while, at the same time, producing errors and mistakes that reinforce human injustice and capitalist domination. These techniques aim to detect and classify patterns in massive groups of numbers, commonly referred to as *big data*. The purpose is clear: creating rankings, standardizations, population comparisons, better customer profiles, and new kinds of "others" that can be observed, targeted, and controlled.

It is no coincidence that the discovery of new patterns used in the new numerical sorcery we call *data science* is primarily based on statistical methods. Statistics has long been associated with forms of state power. In the late eighteenth and early nineteenth centuries, when the practice first emerged, *statistics* (*Statistik* in German) was literally "data about the state."[15] Statistics focused on gathering numbers from every imaginable arena of life: birth and death rates, marriage, crime, suicide, and deviancy. In the process, statistical data came to measure the collective behavior of individuals—so-called mass phenomena—to enhance the state's political, economic, and social knowledge and power over their populations. The individual was quite simply dissolved into the mass.[16]

Statistics evolved to become a rigorous mathematical discipline only in the late nineteenth century, morphing from a descriptive tool into something else: a method of scientific and social scientific analysis that used numbers to quantify, and thus control, the social order by deciding who/what would be classified as "normal" and who/what would deviate from the norm.

In the age of big data, statistical techniques take power to the next step—creating targeted individuals by analyzing mass patterns, not only to count and quantify but also to model and predict. These new quantified prediction cultures based on big data are described by scholars with names such as *surveillance capital, dataveillance, knowing capitalism, datafication,* and *extractive capitalism.*[17] In addition to their destructive political and social effects, all of these paradigms also express a new economic model of experiential extraction—one "that captures our behavioral data and experience and uses it for commercial extraction, prediction and sales."[18]

The models are all based on a compelling premise. Our *experience* is the new raw material, a resource like oil, gas, or water that can be converted into numbers and mined for indications of future behavior. These predatory techniques depend on the vast digital information infrastructures possessed mainly by the behemoth FAANG corporations (Facebook, Amazon, Apple, Netflix, and Google): mass deployments of sensors, artificial intelligence, server farms, data centers, marketers, and advertising agencies to capture and shape our behavior.

For these firms this is a win-win situation. Deploying algorithms that reinforce likes and dislikes and preferences expressed in our continuously ongoing social media performances gratifies and satiates us. At the same time, FAANG companies and others get immensely wealthy and powerful by using this data to create highly lucrative "prediction products" that imagine the future—not what we will do, but what we *could* do. Hal Varian, Google's chief economist, cuts to the chase when he writes, "Everyone will expect to be tracked and monitored, since the advantages, in terms of convenience, safety, and services, will be so great. There will, of course, be restrictions on how such information can be used, but continuous monitoring will be the norm."[19]

Nonetheless, the repercussions of surveillance capitalism are staggering: the total transformation of privacy; continuous monitoring, whether for protection or containment; technical systems whose human-led mistakes lead to discrimination along racial, class, and gender lines; the reconfiguring of human emotions and experience as yet another commodity; the possession of newly gathered knowledge and power in the hands of a few high-tech monopolies on a peninsula south of San Francisco and in the glass towers of Beijing. These effects are already in the news. They are the subject of countless books, articles, podcasts, lectures, policy papers, and discussions within governments, corporations, universities, and homes. We ignore them at our own peril.

But is this the end of the story? True, our human senses, the means by which we see, hear, touch, smell, and taste the world, are now effortlessly captured by artificial sensors and turned into machine-readable data to be exploited by a tiny cluster of the world's most powerful and extractive data corporations. Across literature, film, and even religious practices, however, human beings have long had an almost limitless desire to mold, shape, and share their senses, attention, bodies, and selves with machines: for example, Ismail al-Jazari's thirteenth-century treatise *The Book of Knowledge of Ingenious Mechanical Devices*, which was filled with Islamic visions of automation; seventeenth-century Japanese *karakuri ningyō* mechanical dolls that provide a basis for the Japanese fascination with robots; Mary Shelley's *Frankenstein*; visions of intelligent machines and androids in films like *Metropolis*, *2001: A Space Odyssey*, and *Blade Runner*; and the Hindu practice of offering "puja" (prayers) to kitchen appliances, computers, televisions, cars, and factory machines.[20] Isn't there something deeply seductive about the new kinds of relationships that such sensing machines appear to offer?

In other words, is there something more to sensing machines than just stories of (admittedly) powerful corporations harvesting our behavioral data to maximize their profits? Do we not interact with sensing machines because they enable our desires to go beyond our own human capacities? Do we not find solace in how these machines transform messy human experience into the "objective" ranks of mathematics and statistics? Do we not seek new possibilities in the novel forms of perception and sensation that these technologies offer? In other words, it may not be only what sensing machines desire from us. It's also what we desire from them.

This book explores these dual and sometimes contradictory dimensions: how our "sensed self" has come to be measured and engineered but also captivated and enthralled by these technologies. Our interaction with sensing machines is complex: multifaceted, contradictory, and ambiguous. My goal here is to convince you that our sensed self did not just appear overnight because of Google or Facebook's predatory data practices. It arose in a much messier way, from a specific set of historically rooted and sometimes conflicting social/cultural/economic contexts and desires.

In these pages, the sensed self emerges through what scholars call *material imaginaries*: the imagined but also realized concepts, ideas, projects, schemes, and devices of scientists, artists, designers, engineers, architects, and entrepreneurs who have sought new ways for our human senses to interact with the "senses" of machines. Moving through research labs and galleries, theaters and restaurants, and the living room, kitchen, and bedroom, I explore key ideas and the protagonists who, inspired by biological systems and mathematics, computer science, and the arts and design, are developing new relations between human and machine senses and how these imaginaries have created new ways for our senses to be mapped and manipulated, quantified and commodified, discriminated against, expanded, controlled, and tantalized. Along the way, *Sensing Machines* investigates how these imaginaries have produced new visions of entanglement and feedback among our senses and bodies, technologies, and the wider environments in which we live.

Why write this book now? At this moment in history, there is much-needed criticism against technological utopias. That the visions of democratization, liberal values, and new forms of community long championed by the computing and information behemoths have led to human and planetary inequity and disaster on a scale previously unimagined must be reckoned with. As one recent book rightfully claims—your computer is literally on fire!

On the one hand, sensing machines make mistakes and are predatory, wielded by powerful corporate interests that have mainly economic concerns in mind. On the other hand, these machines have created new

Prologue

possibilities and imaginaries between us and our technological "others." In other words, this book tries to straddle the delicate space between debunking and celebrating sensing machines, by taking specific histories, contexts, interests, needs, and desires into account.

At the same time, there is another motivation to understand how these sensing machines act on our world, and it comes from a very practical position. For more than twenty years, I've worked as an artist and researcher. In collaboration with others, I've built, programmed, and used such sensing machines, which gives me firsthand knowledge of how these technologies work. I deployed them, however, for an unusual purpose. I've used sensors and computers to create large-scale experiential artworks that respond to the movement and presence of visitors. The sensors' data then orchestrates actions in environments filled with light, projections and sound, intense vibrations and fog, and other things that immerse and saturate people's senses. These artworks aim to alter the senses of the visitors' bodies, their perception of time and space, and, perhaps most interestingly, their awareness of their selves.

Through years of making and exhibiting these projects all over the world, I became interested in how *sensors* could be used to answer a central question: How can we design expressive responses in different media to transform a visitor's attention and their physical presence in a space in order to achieve an artistic or aesthetic experience?

One particular project, called *Just Noticeable Difference* or *JND*, developed in 2010 and toured to museums and festivals around the world, embodies this attempt.[21] The installation takes its name from a well-known concept in psychology: just noticeable difference (JND), the amount that a sensory stimulus must be changed for the difference to be experienced as a sensation. The JND concept was invented by a nineteenth-century psychologist and philosopher named Gustav Fechner, who will play a critical role in this book.

In the *JND* installation, one visitor at a time, alone in a very small, pitch-black room, lying on their back, gradually experienced minute changes of light, sound, and vibration at different levels of intensity. Pressure sensors that the visitors lay upon picked up tiny, almost unnoticed movements as the visitors quietly responded to lack of external stimuli in the environment by minutely fidgeting, moving, or rolling over onto their sides as if sleeping.

In exit interviews, we received enthusiastic responses. People described feelings of ecstasy, of losing the sense of their bodies in space, and seeing, hearing, and feeling things that were not there. Many visitors described powerful sensations: the sense of "losing" themselves; of blanking out, hallucinating, or, most extraordinarily, not knowing who was experiencing the different sensations. The more interviews we did with different and diverse audiences ranging from cab drivers to teachers, artists, and families in China, France, Portugal, Canada, and the United States, the more we realized that the visitors' perceptions and the sensors, computers, and environment, the machines that enabled these perceptions to take place, were inseparable. Humans and technologies were codependent; they mutually shaped each other.

The turning point in my thinking came almost a year later. I was presenting the project to a distinguished group of scholars who study the interaction between media and science. I had discussed the technical setup and concept behind the installation, as well as the enthusiastic responses of the visitors, when a well-known anthropologist (now a colleague and friend) questioned my motivations. While I implied that the visitors felt liberated because of the sensory experience they encountered, the anthropologist suggested that I was designing or, more bluntly, *engineering* the visitor's senses and that I should perhaps pay more critical attention to what the social and political repercussions of subjecting people to such sensory manipulation might be. Although this anthropologist didn't claim that the work was oppressive, they definitely implied that such sensory manipulation walked an ethical fine line between liberation and control.

Taking this criticism seriously, I embarked on several years of research, reading patents, news feeds, and accounts; doing fieldwork in research laboratories; interviewing experts; and studying cultural and technical histories of how computer scientists, engineers, artists, and designers have imagined, designed, and engineered systems that not only intensify our senses but commandeer them as well. While many academic art and cultural historians, anthropologists, and sociologists have written about the latter phenomena,[22] I discovered there has been much less written about why these technologies arose in the first place and the kinds of imaginaries that their inventors or users aimed to stage and realize. This is what this book explores.

Prologue

The story of *Sensing Machines* comes in eleven short chapters spread across five sections. Each chapter tells a different story of our encounter and interaction with these systems in daily life: playing games, experiencing immersive artworks, driving, eating, exercising, sleeping, and dreaming. Along the way, I draw on examples from computer science, experimental psychology, art history, engineering, anthropology, the histories of technology, science, and psychology, and other disciplines to demonstrate that none of what we are experiencing today is without precedent.

Chapter 1 bounces back and forth between two epochs: the mid-nineteenth century, when the first efforts to quantify the senses through mathematics and machines took place in newly emerging psychology laboratories, and the twenty-first, in which those nineteenth-century techniques are still being used to reimagine and design the latest virtual reality (VR) machines. Chapters 2 and 3, the Playing section, focus on the development of sensor technologies used in music and gaming, in which human bodies were reconceived as new interfaces for expression and play. The two chapters in Immersing, chapters 4 and 5, delve into the use of sensing machines in the arts, from artists' deployment of them in the "programmable art" environments of the 1960s and the immersive installations of collectives like teamLab in today's "experience economies" to a new horizon of autonomous sensing machines that will cooperate with human artists in the not-so-distant future. The Engineering section, with chapters 6 and 7, explores new areas in which our senses are both engineered and intensified—driving and eating. Chapters 8–10, the Monitoring section, open up the black box of sensing machines, examining tracking, pattern recognition, and prediction, as well as the ways in which the technologies of sensors tied to computers are both ubiquitous and hidden, distributed in networks across time and space. Finally, chapter 11, in Enhancing, investigates the larger question of our sensed self: how we attempt to enhance and extend our bodies, minds, and souls through new wearable sensing devices, from brainwave detectors to machines that sense and even shape our deepest sleep and dreams.

Given that this book was completed in the midst of a global pandemic that has had massive political/social/economic/technological repercussions,

I conclude by speculating on the heightened role that sensing machines played during the extremes of the crisis in 2020–2021—one that may potentially bring about a different understanding between ourselves and the technological others that surround and act on and with us.

1 Measuring Sensation

> When the eye ceases to see, the ear to hear, and the sense of touch to feel, or when our senses deceive us, instruments are like a new sense of an astonishing precision.
> —Étienne-Jules Marey

One day in the year 1840, a man opened his eyes and couldn't see. This was it, the "final blow," as he later wrote in his diary.[1] It was as if the man, a renowned German medical doctor turned professor of physics, had inexplicably gone blind overnight. But his condition was not new. It was the dramatic culmination of months of unexplainable symptoms that had befallen this scientist: bursts of light in the eyes, headaches, nausea, lack of appetite, insomnia, and neurosis. Little did the scientist know, however, that his dire situation would eventually result in something remarkable—a startling revelation that would forever change our understanding of the human senses and how they would come to interact with machines.

Between 1839 and 1843, Gustav Fechner, a son of a protestant minister and a trained physician who would soon become a major force in the emerging sciences of psychology and physiology, suffered from a mysterious "malady." The illness had a striking effect on Fechner's work and life. He lost interest in conversing with others. He couldn't see properly, with his eyes continually overwhelmed by flickering artifacts and erratic flashes of light. His head throbbed with dull pain as he wandered aimlessly inside his bedroom, study, and sometimes garden for almost three years, self-isolating and wearing a handmade mask of lead cups that protected his eyes from the blinding daylight. He even painted his bedroom black to stop light from leaking in.[2]

Fechner's eyesight wasn't the only thing that suffered. A growing lack of appetite stretched over months, reducing him to a near skeleton. He stopped speaking in eloquent phrases and sentences. His attempts at healing himself through "animal magnetism" (hypnosis), homeopathy, electric current, and moxibustion (the burning of herbs near the skin) were to no avail.

No one quite knew why Fechner fell ill. Burnout? Too much work, like partially writing and editing a seven-thousand-page, eight-volume encyclopedia? Or turning himself into a human guinea pig in the name of science, damaging his eyesight? Fechner had stared too long into the sun using glasses with only colored filters as he explored the perceptual phenomena of afterimages—the images that stay on the retina long after one stops gazing at a light source. This series of experiments seemed to throw him into a searing, never-ending "light chaos" that he would constantly experience, even with closed eyes.[3]

"Close to insanity," Fechner nevertheless began to slowly recover from his malady. Instead of gradually adjusting his eyes to faint light, he took the brute force route: sudden and intensive short-term exposure to the brightness of the everyday, quickly closing his eyes before the light caused intense pain. He resumed eating, consuming such odd delicacies as raw ham soaked in wine and lemon juice, as well as sour berries and drinks. Although he still experienced "disagreeable sensations" in his head, he finally spoke again.[4]

One October afternoon, Fechner wandered into his garden as he occasionally had done during his illness. This time, however, he took a gigantic step to reintegrate into the visual world. He removed the thick bandages covering his eyes. The light spilled in. As he glanced into his garden, the scientist experienced a miraculous sight. He saw the flowers "glowing." They seemed to speak to him. In this ecstatic moment, Fechner came to an astonishing realization—plants must also have souls.

Fast-forward 180 years. In the digital haze of pandemic newsfeeds, you are clicking through pages on LinkedIn. Dozens of jobs in new professions with strange sounding titles appear: vision engineer, applied perception scientist, visual experience researcher, color scientist, and neural interface engineer, the job description of which is to "help us unleash human potential by eliminating the bottlenecks between intent and action."[5]

One career in particular catches your eye: an applied perception scientist, working for Oculus, a once-small start-up that manufactured a lightweight VR headset, which Facebook bought in 2014 for $2 billion dollars. The job announcement asks for expertise in visual perception, the "computational modeling of vision," and "experimental and/or modeling approaches" that "help us inform AR/VR display requirements and architectures."[6] This new career in applied perception science also has another thing in common with the other LinkedIn jobs—it asks for knowledge in an obscure sounding discipline called *psychophysics*.

What does a scientist undergoing a mysterious illness in mid-nineteenth-century Germany have in common with twenty-first century engineers seeking to plumb the depths of human perception? In 2020, Gustav Fechner, physicist, philosopher, and believer in the ever-lasting consciousness of souls, plants, and the earth itself, is a forgotten figure. But he shouldn't be. Fechner is one of first scientists to propose an idea far ahead of its time, one that has had a radical effect on how we view sensing and perception in relationship to man-made machines.

Gustav Fechner asserted that we can measure and calculate how we sense the world using mathematics. Indeed, in the early morning hours of October 22, 1850, just seven years after his malady subsided and the encounter with the flowers in his garden, Fechner had another burst of inspiration. He came to the realization that there must be a relationship between spiritual

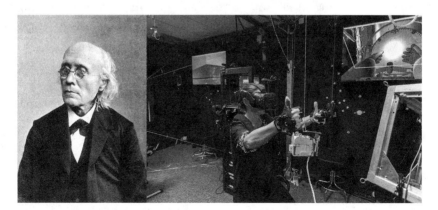

Figure 1.1
Left: Portrait of Gustav Theodor Fechner (circa 1883–1884). Artist unknown. *Right:* NASA training in VR Lab, 2018. Photo by NASA/Robert Markowitz.

and physical energy, *a measurable correspondence between the world external to our sense perception and the internal world of our brain processes.*

But Fechner needed to prove his theory scientifically. He thus invented the almost mystical-sounding discipline that he christened *psychophysics*—a "theory of the relations between body and mind." In Fechner's formulation, psychophysics would be an "exact science, like physics" and "rest on experience and the mathematical connection of those empirical facts that demand a measure of what is experienced."[7] It aimed at no less than to establish a measurable connection between two spheres that had long remained separate: the material, physical universe and the mental, psychological one.

Psychophysics set the European scientific world on fire. It helped advance the newly emerging discipline of experimental psychology, in which there was already a mad rush to translate human thoughts into numbers. The rising hybrid scientists of the period—psychologists, philosophers, mathematicians, and physicists—were eager to escape a nonscientific (e.g., unmeasurable) understanding of how the senses and the mind worked, and Fechner supplied them the ammunition. These scientists began to develop theories to demonstrate mathematical connections between physical phenomena, what are called *stimuli*, and the sensory experience of such phenomena, labeled *sensation* or *perception*. But in the process, they also sought to eliminate the experiencing, subjective self doing the sensing, replacing human sensory experience with "objective" formulas and equations.[8]

Fechner's ideas would also quickly be materialized in the newly appearing sensing machines of his time—instruments with strange sounding names like *kymographion, tachistoscope,* or *chronoscope* and which measured blood pressure; the speed of vision; or response time, the period it would take for a person to react to a stimulus. In the words of the nineteenth-century French physiologist Étienne-Jules Marey, a major inventor of such devices, these new instruments sought to reveal the hidden "language of nature."[9] Such human sensory measuring devices were to be found in a novel kind of experimental scientific environment: the emerging experimental psychology laboratories in Europe and the United States, whose goal was to create a new kind of human being: quantifiable, calculable, and predictable.

We would assume that psychophysics died a dusty death, relegated to the history books of psychology and the crumbling sets of abandoned scientific instruments that fill up university collections. But this assumption is an error. As our LinkedIn search reveals, psychophysics is very much alive in the most unimagined of places: the labyrinths of behavioral research at Facebook Reality Labs, the game testing cubicles of Electronic Arts, or the perception laboratories of various universities, who are united around a similar aim. They use the sensing machines of our time—networks of sensors, statistical modeling, machine intelligence, and computing infrastructures, human labor and the earth's resources—to capture, calculate, model, and simulate human sense perception beyond the wildest dreams of nineteenth-century scientists and, in the process, create a wholly new relationship between these sensing machines and us. In fact, Fechner's rendering of sensory experience into numbers has won out. His belief in the life of plants, the earth, and the cosmos itself now includes new entities that we interact with and live among: machines that sense, act on, and perceive our world.

Why is a nineteenth-century science that measured how we perceived the world still relevant today? As a natural scientist, Fechner revolutionized different areas of study. With his successful research and experiments in electrophysics and electricity, he founded the first institute and scientific journal dedicated to physics in Germany. Despite his belief in the physical basis of things, however, Fechner was still uncomfortable with the dominant philosophy of the time—what was called materialism, which argued that reality only exists because it is reducible to mechanical laws. What you see is what you get and nothing beyond.

Instead, Fechner sought a unification, a linkage between things that philosophers and scientists had traditionally kept apart: mind and body, material stuff and immaterial consciousness, even life and death. This linkage is what he called the animated substance of the world. Because Fechner did not separate consciousness from physical matter, his view of the world could be described with the philosophical concept of *panpsychism* (literally, *all souls*): that the soul (from the Greek word *psyche*) is rooted in everything, from rocks and minds to plants and stars and finally the earth itself. To panpsychists, the material world is alive and even conscious.[10]

Perhaps the ultimate contradiction (at least for a natural scientist) was Fechner's impassioned belief in immortality. He entertained interests in the afterlife and parapsychology, even attending séances for the dead. Yet even this was not something considered esoteric or strange like it might be today. Such interests were common, even among scientists, during this period.

It was in Fechner's voluminous and obscure writing that these seemingly outlandish views found their home. In dense books running hundreds of pages long, with enigmatic titles like *Nanna: The Soul Life of Plants* (named after the Norse goddess of flowers Nanna) or the epic *Zend-Avesta: On Matters of Heaven and the World Beyond*, Fechner spun out philosophical beliefs in which spirit, soul, body, and nature were densely connected in an interlocking web.

Fechner laid out two different visions of the world: what he would call the *day view* and the *night view*. The day view was comprised of mind and spirit and encompassed Fechner's anti-materialist beliefs in the aliveness of all things regardless of whether they would be considered biological organisms.

The night view foretold the opposite: the mechanistic and materialist world. These two worldviews were crucial for the *Zend-Avesta* because it was in that book that Fechner also laid the foundations for a new kind of "mathematical psychology," in which a relationship could be drawn between *stimuli*, consisting of material phenomena, and *sensation*, consisting of psychic or mental phenomena. "There is nothing," Fechner wrote, "to stop us from considering the materialist phenomena that underlie a given psychical event as a function of the psychical event and vice versa"[11]

The goal of Fechner's new mathematical psychology was unambiguous. It sought to develop a rigorous, quantifiable science that would replace the fuzzy, speculative understandings of mental phenomena that had already gripped the emerging discipline of psychology. But Fechner's mathematical psychology would do something more radical. It established a quantitative, rigorous relationship between matter and mind, forging a new connection between the accessible material world and the inaccessible spiritual one.

In 1850, Fechner didn't know exactly how the relationship between stimuli and sensation actually worked. But he had a hunch that a measurable connection could be determined between the two.

He proposed a simple question: How could one *measure* sensation and, therefore, perception? Fechner didn't start from scratch in this formulation.

Instead, he drew on an existing theory from one of his contemporaries, German psychologist Ernst Heinrich Weber, who had put forward the idea that there was a relationship between the intensity, *the strength of a stimulus*, and its resulting sensation.

Imagine the following scenario. You have two equally weighted containers. By lifting them, you compare how heavy they are. Then, someone adds a little more weight to both vessels and asks you to verbally state if you feel a difference between the two. The trick is that you will only notice a change if you sense that the *difference* is large enough. Your task is to determine how much the weight changes in order to distinguish one from another.

This scenario is not just dreamed up. Weber, and later Fechner himself, actually tested it. While researching the sense of touch, Weber came up with a concept called the *two-point threshold*. Using a metal compass touching the skin of a test subject, Weber asked what the smallest distance between the two points of stimulation would need to be for the subject to report them as two distinct points.[12] He called this measurement the *difference threshold*—the minimum amount by which the intensity of a stimulus would have to be changed for the subject to perceive a difference in their sensory experience of that stimulus. Revealed in both the weight comparison and the two-point threshold experiments, the difference threshold also went by another, now more famous name: *just noticeable difference.*[13]

Fechner had mathematically restated what he called *Weber's law*. He demonstrated that while sensation was a function of a stimulus, there was not an assumed one-to-one relationship between that stimulus and its perception. In other words, one's sensory response to a stimulus was not proportional to the physical intensity of that stimulus.

Weber's law had shown in an intuitive (but not quantitative) manner the amount that a stimulus would have to change before a subject would recognize an experienced difference in the change of the sensation. Through a series of calculations, Fechner transformed Weber's concept into a mathematical formula, arriving at an equation which expressed the ratio of the JND of a stimulus to the stimulus itself.[14] Fechner then manipulated Weber's law into a formula that would later be called *Fechner's law*, in which he claimed to *precisely show the relationship between the mental and the physical*—the mind and the body: $S = k \log R$, where S = the intensity of the sensation and $\log R$ represented the intensity of the stimulus, with k as a constant.[15] To put this in a nonmathematical way, as a stimulus increases in intensity, the intensity or *magnitude* of the change of sensation also has to continually increase in order for us to perceive a difference.

Although set into mathematical terms by Fechner, his equation of S = k logR closely resembled another similar and well-known mathematical formulation: Ohm's law—which also expressed a relationship between two variables (this time, voltage and current) in logarithmic terms.[16] This relationship between concepts in electrical behavior and human perception was not coincidental. Like many nineteenth-century scientists, Fechner was obsessed with the concept of energy and could thereby equate the energy inherent in pure physical phenomena with that of bodily phenomena.[17]

The concept of energy had an almost mystical aura surrounding it. Fechner's formulation of his *fundamental psychophysical law* was focused on the energy inherent in a stimulus and the resulting energy contained in the sensation of that stimulus. In fact, when Fechner had his epiphany about psychophysics, he immediately drew on these energy theories, recognizing that "the relative increase of bodily energy is related to the measure of the increase of the corresponding mental intensity."[18] At its basis, *psychophysics was about the process of measuring difference in the energy of a stimulus and when that difference would become noticeable in perception.* A law of mathematics was thus translated into human perceptual terms.

While all this might sound needlessly complex, a common example illustrates the basic principle of Fechner's psychophysics. Imagine a standard hearing test in which the volume (magnitude) of a sound is increased or decreased. The person running the experiment asks you to verbally report when you hear something and begins to adjust the volume. In other words, the experiment asks you to identify the exact moment when a tone that at first you cannot hear suddenly becomes audible—perceivable as an auditory sensation. At first, you report that you hear nothing; the volume as a stimulus is too quiet for your ears and auditory nervous system to perceive. Gradually, however, you seem to hear the tone; it can be consciously detected.

Fechner had a name for this sudden moment at which a stimulus could be detected—he called it the *absolute threshold*. The absolute threshold describes when the intensity of the stimulus "lifts its sensation over the threshold of consciousness."[19] The absolute threshold could thus measure the smallest amount of stimulation that could be detected by an organism.

But the absolute threshold is only one value on a longer-intensity scale. Fechner therefore had to develop another measurement that would consider the change of perception as a stimulus would become more or less intense. To put this another way, what would the just perceivable or just noticeable *difference* be as the intensity of the stimulus strengthened or weakened over time (figure 1.2)?

Although detailed over hundreds of pages in more complex numerical analysis, Fechner's psychophysical principles revolved almost entirely on this concept of perceivable difference in the relation between stimulus and sensation/perception. The same could be said for the three core methods that Fechner introduced as core psychophysical principles still in use today—namely, (1) the method of limits or JNDs; (2) the method of adjustment (average error); and (3) the method of constant stimuli (right or wrong cases).[20]

⌒⌒

You don the cumbersome virtual reality headset (figure 1.3). Once it sits on your head, the device is turned on. At first, there is a menu that instructs you how to use the handheld controller, which allows you to navigate the visual space soon to be displayed before your eyes. Pressing the left button allows you to move left; the right button, right. If you press the up and down buttons, you travel vertically inside the animated world. After this episode, known as onboarding, the experience begins.

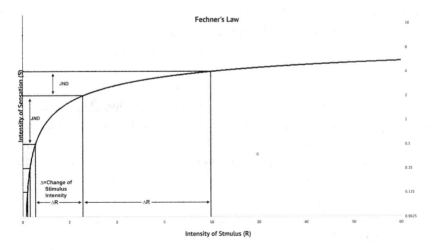

Figure 1.2
Weber's (later called Fechner's) law.

Figure 1.3
VR at Samsung Mobile World Congress 2016.

You are under the sea, accompanied by hundreds of animated swimming creatures in all kinds of fluorescent colors. Turning your head too quickly, you suddenly feel a spell of dizziness. As you navigate this animate and animated undersea world, you run into corals and rocks on the canyon bottom of the ocean floor. Pressing the buttons on the controller doesn't really help, so you just hang out. The undersea world drifts by.

What's more amazing is that this artificial thing, firing complex and detailed 3D images into your retinas, binaural sound into your ears, and vibrations through your hands, follows your movements and actions. As you rotate your head, the sound follows, creating a sphere of audio around your skull. Some of the swimming sea creatures almost bump into you, and sometimes they make contact, bouncing off of you as if you were really there under the sea. Sometimes it seems that the space projected in front of your eyes and into your ears is infinite in dimensions. Other times it feels confined—as if you were restricted to moving only a few steps before reaching the edge of the animated world.

But this is all a ruse. The VR experience you undergo is entirely artificial. None of what is happening seems remotely like how the physical world that you walk in or move through daily actually functions. In the "real world," you have to walk toward something to hear it, and, in fact, the sound source that is closest to your ear is the one you hear first. Similarly, if you want to look at something to get a closer sense of its detail, you have to literally move your body toward the object. The object rarely moves toward you (even if it's alive) in order for you to get a closer look.

How then is this perceptual trickery accomplished? Although the Oculus or HTC Vive *head-mounted display*, as the technical term goes, seems like a glorified but worn projection screen, it is anything but. This roughly 470-gram display is packed with OLED screens and sensors that can measure an inconceivably large range of physiological data: the speed of your eye movements (called *saccades*) or when you blink; where your head is located in space relative to the image you are looking at in the headset; how steady you can hold your hand as you grip a wireless controller or navigation device; or where your ears are in relationship to the visual scene. It's important to realize how critical these sensors are for guaranteeing the holy grail of a VR experience: creating a sense of presence by being physically immersed in a "nonphysical world."[21]

Sensing research for VR and for augmented, mixed, and extended reality (AR, MR, and XR) develops in leaps and bounds. As of 2019, so-called six degrees of freedom (6DoF) sensors were the latest technology to be integrated into VR headsets. Based on the engineering concept of *degrees of freedom*—the number of directions a rigid body can move in 3D space—these sensors enable the real-time tracking of full body rotation and position: forward; backward; up and down; left and right; and rotation around the three X, Y, and Z axes, referred to as pitch, yaw, and roll. In essence, any movement you make can be captured and organized by these sensing devices.

While the ability to track such parameters previously relied on external "outside in" sensors, such as cameras or lasers that are placed in a stationary location,[22] the integration of "inside out" sensors, ones placed directly into headsets, enables a new level of simulated experience. The ability to move around in a virtual space will thus soon parallel the way one moves in the real world.

But what does Fechner's psychophysics have to do with VR? More than we can at first imagine. The detection and discrimination of limits, errors, and thresholds—the basic tools of psychophysics—are one of the fundamental scientific methods used to test these new reality devices against your own way of perceiving the world. As one group of cognitive and computer science researchers claim, "VR can be seen as a continuation of a long psychophysical tradition that attempts to interfere with our perception in order to clarify its underlying mechanisms."[23]

Over the more than one hundred years since Fechner's invention of psychophysics, the discipline has advanced. It is no longer only used to measure the complexities of human perception to gain knowledge about the human senses. Psychophysics has become a *design method* for creating sensing machines and the artificial experiences that such machines make possible. Sensors together with psychophysical methods are not just measuring the world, they are helping create alternative ones. Within Facebook Reality Labs, with its almost military secrecy, scientists with PhDs in neuroscience, applied perception research, robotics, and computer science still draw (albeit with updates) on the quantitative modeling of sensation, stimuli, and perception that Fechner discovered in the late nineteenth century in their twenty-first century aims to create VR, AR, and XR experiences that are both exceedingly real and, at the same time, completely artificial.

In applied perception research, machine sensing, and psychophysics, measuring human sensing and perception go hand in hand. Take one of the major issues in VR: calculating the position and orientation of a user's head in space to dynamically adjust the image in each eye in order to mimic the stereoscopic way we see. If there is *latency*—a temporal lag between an input action such as a head movement and the resulting image in the visual display—then the sense of virtual presence is disrupted. Worse, this temporal delay can lead to motion sickness or the even more bizarre experience of *oscillopsia*—the perception of a moving image even when the image is stationary.[24]

To measure the difference between action and visual response, researchers use modified versions of Fechner's method of limits, in which a changing stimulus (in this case, a moving object in a virtual environment) is presented at various intensities together with a standard, constant stimulus to determine whether or not the range of intensities is the same as the constant. Calculating these JNDs enables vision scientists to understand whether or not their test subjects could, in fact, perceive just noticeable differences between their head movements and the speed of movement of a virtual object due to the updating of the 3D image.[25]

Psychophysics is also used as a design method in VR/AR research when determining how fast one's pupils dart about when immersed in a virtual scene—what is called *saccadic eye movement*. One technique called *redirected* or *infinite walking*, freshly emerging from computer science research, utilizes our rapid eye movement (REM) or saccadic suppression that we make when gazing at a scene to actually trick the eyes and brain into thinking we are in a virtual space that is larger than the actual physical space we are in.[26]

Saccades produce a kind of microblindness, a split-second period when the eye actually is closed but we don't perceive it. By measuring the change in rapid eye movement using internal gaze-tracking cameras built into a VR headset and "incorporating guided navigation and planning based on the scenario," researchers take advantage of this momentary gap in seeing that occurs during saccadic movement, "redirecting the user much more aggressively, yet still imperceptibly." This works as follows. Eye trackers calculate the length of the saccades and, within those microintervals, make small, imperceptible adjustments to the rotation or position of a virtual camera in the virtual space so as to "exploit as opportunities for imperceptible transformation of the world."[27] Sensing our eye movements serves to fool the eyes and the brain into believing things have not changed in the virtual scene when they actually have. By then measuring a range of JNDs, the researchers aim to understand how perceptible such changes are to the viewer—or if they are barely noticeable or not noticeable at all.

The complexity of VR and AR headsets and goggles, however, makes them increasingly more like airplane control dashboards than something meant for the living room. The inner guts of the Magic Leap—an MR goggle-based worn device that appeared in late 2018 after many years of stealth development and hype—is covered in so many sensor arrays, from IR-driven eyeball tracking to 6DoF sensors, that its sensing infrastructure has been compared to that of a self-driving car's (figure 1.4). Indeed, in an interview, the former founder of Magic Leap makes clear this strategy of direct perceptual integration: "Your brain is the coprocessor. We spent all of our money and everything we're doing to send a signal for the human brain, not a camera CCD [charge-coupled device]. Not a monitor. Not anything else. I came out of biomedical engineering and the idea was, don't break the brain. Don't break the brain is the number one rule."[28] It seems

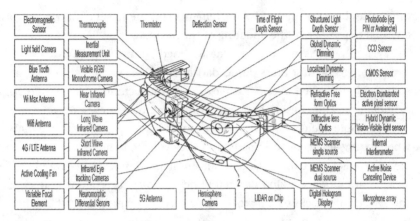

Figure 1.4
Magic Leap sensor architecture. Drawing from original patent.

the Magic Leap aims to leave no potential body interaction (with the device) unsensed.[29]

This desire to sense everything is not just technological. It is psychological as well, part of a larger movement to reorganize the human senses as input for designing the simulated immersive worlds that Facebook, Apple, Samsung, Microsoft, and a myriad of Chinese start-ups hope many of us will soon inhabit. Because the key to VR, AR, MR, and XR is achieving a degree of absolute believability, what in the theater has long been called *willing suspension of disbelief*, from a psychophysics perspective it is necessary to numerically judge how much believability of presence an artificial world conveys so that eventually users can no longer tell the difference.

This total synthetic belief is also evidenced by the shift in the rhetoric around VR itself. Long accused by philosophers and cultural critics of denying our bodies, now VR can't seem to get enough of them. In fact, what used to be termed *virtual reality* is increasingly described as *real virtuality*.[30] In a seemingly slightly desperate attempt to integrate its users' bodies into its simulations, VR has been recast to create a believable sense of bodily presence that ironically can only be achieved in the virtual world by artificial perceptual machinery: sensors, high-resolution displays, and computers processing millions of high-polygon graphics.

Contrary to the idea that the senses are simply to be replaced by artificial sensors, a different story is emerging. Perhaps more than ever, our senses are needed to feed ever newer immersive experiences by increasingly being what researchers call *tightly coupled* to these devices—an inseparable connection that can be achieved by psychophysically measuring the changes in stimuli produced by these systems and incorporating those changes into actual design requirements for the hardware and software itself. Yet, our bodies and senses that carry different histories—cultural, social, economic— are tabula rasa for these systems. Gender, class, race, or ability differences are simply erased in favor of the psychophysical norm.

Modern perception science is fundamentally based on these principles of cultural sameness. But psychophysics is not just erasing cultural and social distinctions between bodies but also creating new loops between design and perception. By providing recommendations for the hardware and software design of new technologies that enact researchers' theories about how our sense perception works, these researchers can then create perceptual experiences that reinforce these models, regardless of the bodies that are part of these new realities.

Today, the relationship between sensing machines and sensation and perception seems a given. But unlike the sensors inside the Oculus Quest or Magic Leap that can instantiate whether a perceivable change happens in the frame rate of an image, Gustav Fechner in 1860 had little access to sensing devices to experimentally prove his theories. His "sensors" were cruder: the perceptual abilities of human beings who, under psychophysical tests, would generate verbal data about what they experienced, which then could be calculated to come up with measurements.

In other words, as mathematically rigorous as they were, Fechner's psychophysical methods still relied on human scientists, physiologists and psychologists who would "subjectively" report what they had perceived from test subjects during an experiment. The experimenter could not control whether the subject's report would be correct or even accurate.

To complement Fechner's psychophysics, nineteenth-century scientists therefore turned to newly emerging technologies to better measure sensorial responses: new sensors to prove their new theories. These researchers invented instruments designed to capture and measure the human (and also animal) senses. From the ophthalmoscope to the acoustic whistle, the olfactometer, chronoscope, aesthesiometer, and photographic gun that enabled the emerging practice of chronophotography, these instruments became, in effect, de facto senses.[31]

Not only were they the earliest versions of the sensing machines on offer today, but these sensing instruments also played a fundamental role in the construction of a vast new domain of knowledge about the human sensorium called *sensory physiology*, which understood the senses as key to the development of psychological and physiological knowledge.

Utilizing scientific observation, experimental procedures, and emerging instruments, sensory physiologists studied a broad range of phenomena, including spatial perception in hearing and seeing and the speed of neuronal firings or sensory *quanta*: tiny measurements in the form of thresholds and differences of stimuli intensities, mainly derived from Fechner's psychophysics.

Sensory physiology took the body and the senses directly into the technological loop; there was not merely a chance or accidental relationship with the technologies of measurement and analysis that would soon proliferate in the first research laboratories dedicated to experimenting upon and analyzing the senses of living bodies.[32] Not only were the senses reconceived as technologies in and of themselves, but also, like our VR and AR headsets, instruments became increasingly integrated into animal and human senses. In other words, the senses became sensors, and sensors assumed the role of sensing.

This "extension" of the senses into instruments seemed par for the course in the early to mid-nineteenth century. Already devices such as the stethoscope and the thermometer were replacing the human senses. The overall effect of these new experimental technologies and the laboratories where they were deployed was that machines not only increasingly regulated the bodies and senses of the subjects being studied but also shaped the senses of researchers themselves. In other words, researchers became data analyzers.

Indeed, for those scientists working in the shared space between physiology, psychology, and medicine, instruments became essential partners in revealing the invisible forces fluxing through bodies, forces that were inaccessible to the human senses. There is no clearer expression of this sentiment than the words of nineteenth-century French physiologist Étienne-Jules Marey, who stated, "How little our senses tell us, so that we are constantly obliged to use apparatuses in order to analyze things."[33]

Three characteristics marked this era of early sensing machines. First, the senses (and sensing itself) became increasingly equated with measurement. For instance, through his experiments with frogs and human muscles on *reaction time*, the temporal difference or lag between a stimulus and its response, German mathematician, philosopher, and sensory physiologist Hermann von Helmholtz introduced new empirical techniques for separating and then studying the individual senses. Helmholtz's reaction-time experiments proved that sensation and quantification were intimately linked with one another. A sensation applied to a muscle in a frog or the skin of a human would naturally result in a reaction, quickly followed by a measurement of that reaction.[34] It's almost as if the experiments were designed to demonstrate the efficacy of the technologies themselves.

Second, such experiments could not have been accomplished without specially designed instruments that enabled the measurements in the first place. As physiologist Marey made clear, such instruments were needed to capture temporal changes that outstripped the human senses' abilities to perceive them. Because sensory experience was simply too slow, the precision of measurement became a core goal.

Third, physiologists, psychologists, and physicians sought to turn the messy, imprecise senses into something externally readable through an early form of data visualization—what Helmholtz and Marey called the *graphic method*.[35] The capturing of sensor data and its writing or *inscription* onto a surface was pioneered by German physiologist Carl Ludwig, who, in 1846, introduced a new machine called the *kymographion* (literally, *wave writer*)—an instrument designed with an explicit graphical purpose: to trace the shape of the heartbeat.

The machine worked through the insertion of a small air-filled bulb into the artery of a live animal. A change of blood pressure would cause the bulb to move up and down, and this motion would subsequently be transferred to a stylus mounted against the head of a rotating drum. As the ink-based stylus contacted the drumhead, visual tracings of the pulse would appear on its surface.

The kymographion was an early device to capture physiological data as visual traces in time, and the graphic method took advantage of this by recording time intervals that escaped the human eye's temporal and spatial resolution.[36] For Marey and Helmholtz, who both pioneered the graphic method around the same period, the possibilities of visualizing hidden forces opened up a new chapter into the analysis of the human body's inner life.

This early method of data visualization also suggested something else: the removal of the human observer from the scientific process of perception. There would no longer be a human intermediary between an action and its visual representation. In the blossoming world of nineteenth-century psychology, sensation could thus only be justified if it could be calculated and visualized. Many years later, American designer Edward Tufte reinforced what many nineteenth-century scientists who justified the graphic method were arguing for when he gave a new name to this interest in the visualization of data: the *visual display of quantitative information*.[37] The pioneering communications designer Muriel Cooper, the first tenured female professor at the MIT Media Lab, went further—she dubbed such quantified visual data "information landscapes."[38]

⌒

This age of quantifying living bodies took off. But it increasingly demanded ever-stranger sensing machines to advance its scientific cause: primitive electrodes; pneumatic tubes and mechanical harnesses that could be attached to the limbs, arms, wings, feet, and legs of unfortunate humans and animals, such as Marey's air pantographe, which was used to study live birds in flight (figure 1.5); recording devices like pneumographs, which graphically represented throat movements produced during vocalization; or tachistoscopic apparatuses that measured how visual sensory impressions could affect consciousness within specified time intervals.

Quantifying the senses also required new kinds of infrastructures in the form of sophisticated spaces, rooms, and human resources (i.e., students) who would operate these new sensing machines so that sensory data could be recorded, studied, and stored. Founded by a Helmholtz protégé named Wilhelm Wundt, one of these new infrastructures opened its doors at the University of Leipzig in 1879.

Wilhelm Wundt has a particularly important position as one of the founders of modern experimental psychology. Influenced by psychophysics, Wundt sought to test Fechner's techniques experimentally, subjecting sensory experience to rigorous methods concurrently being developed in sensory physiology and "physiological psychology." Here, the inner world of a subject met head on with external instruments that would scientifically

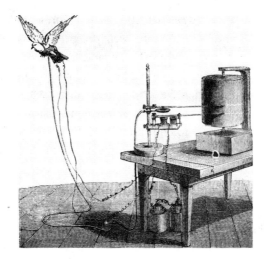

Figure 1.5
Étienne-Jules Marey, air pantographe (machine for studying live birds in flight). In *La Méthode graphique dans les sciences expérimentales et principalement en physiologie et en médecine* (Paris: G. Masson, 1878).

validate their deepest psyche—something that psychology had not yet explored in an experimental, technologically instrumented way. Wundt sought to externalize inner perception, thoughts, and even memories by measuring a person's physiological characteristics: breathing, speed of reaction, pulse, and nerve responses.

Like Fechner, who was some thirty years older, Wundt studied sensation and perception in hopes of being able to make a link, a *contact point*, as he called it, between the physical and psychological. He would succeed in doing this by subjecting his laboratory test subjects to artificially generated sensory experiences in experiments designed to observe and analyze their response.[39]

In Wundt's Leipzig-based Psychological Institute, one of the first experimental psychology research laboratories, this goal would be made possible by the newest scientific instruments and environments: a room painted completely black for vision experiments; isolation spaces with padded doors for acoustic experiments;[40] a space housing a gigantic Meidinger battery that could distribute power to a range of testing apparatuses across different spaces; tuning forks; machines to split and measure the color spectrum; chronographs and chronoscopes for reaction time experiments; and "time sense" apparatuses to calculate the mental representation of time. All of these machines and infrastructures were designed to investigate the key elements of Fechner's psychophysics: to enable the objective measurement of the intensity of sensation produced by the new instruments themselves.

But Wundt went further. He made physiology itself into a new kind of exploratory playground. Wundt's experiments usually required three things: a test subject, instruments, and researchers who could administer, gather, read, and analyze the data. Experiments ran the full gamut: psychophysical tests that measured the quality and intensity of sensations; tactile and auditory psychology; experiments on visual sensations, and the sense of taste and smell; visual depth perception; studies of the time sense and attention; and processes of association and memory.[41]

Unlike previous researchers in the senses, Wundt's laboratory established a collective experimental atmosphere for the training of students in research, one of the reasons that budding psychologists from all over the world came to study with him.[42] **Sensory physiology thus not only grew** through the development of sensory instruments but also through human experimental teams who would administer and analyze these experiments. The laboratory became a teaching environment in addition to a site of new knowledge about the senses.

The laboratory also helped create a new division of labor between the test subject and the researchers themselves. The effect was that the subject became a data source and the researcher an experimental manipulator.[43] But the results were not entirely in the hands of the experimenters. The test subjects themselves were encouraged to develop an experimental and objective understanding of what was happening to their own selfhood at the time of the experiment—what Wundt called *experimental self-observation* or *introspection*. Observers would be exposed to standard repeatable situations and then requested to respond in a quantifiable manner.

What was critical for Wundt's method was the scientific governing of his test subjects, enabled by the experimental laboratory environment. Like the verbal reporting of psychophysical limits and thresholds from Fechner's early test subjects, the controlled conditions of an experiment could allow subjects to immediately report verbally on their perceptions.

Unlike Fechner, however, Wundt verified these reports through measuring instruments. Thus, the subject's own description of what was happening to them, what Wundt called their *inner perception*, might begin to

Measuring Sensation 35

approximate the conditions of external observation. Externally produced forms of sensation here served to influence and shape subjects' sense perception; the more the subjects were subjected to such sensation, the more experienced their observations became without regard to any kind of superfluous self-reflection.

The sophistication of Wundt's laboratory was that in addition to these larger machines, the institute also featured many smaller instruments, some of which could be directly attached to the body in order to gauge the relationship between physiological data (like blood flow, breathing rate, and pulse) and emotion—the subject's response to external stimuli—while the test subject rested in an "indifferent frame of mind."[44]

These new sensing instruments became so essential for rendering new knowledge about human beings that E. B. Titchener, one of Wundt's students who later set up one of the early psychology laboratories in the United States at Cornell University, put it thus: "The experimenter of the early 1890s trusted, first of all, in his instruments; the chronoscope and the tachistoscope were—it is hardly an exaggeration to say—of more importance than the observer."[45]

One specific instrument in Wundt's lab bears brief description. The plethysmograph was a new apparatus designed by the Italian physiologist Angelo Mosso in 1874 to measure the change in volume of an organ based on the shifting blood pressure flowing through it (figure 1.6, left). Consisting of an arm-sized glass vessel filled with water, when a subject put their arm into the tube, the change of blood pressure displaced the water level and transferred this to a stylus, which would graphically record the fluctuations on the surface of a turning drum.[46] Presciently, albeit in an updated, electronic version working with light, Mosso's sensor would wind up some 140 years later in a device attached to the body that also attempts to link hidden interior data to our visible actions: the Apple Watch (figure 1.6, right).[47]

Why did these nineteenth-century "engineers of life" believe that their technologies were infallible?[48] That their graphical curves and markings made "without recourse to the human eye or hand" would tell the truth over the fallible human senses?[49] Despite Marey's belief in such instruments rendering visible the hidden language of nature, these instruments were eventually criticized as imprecise.[50]

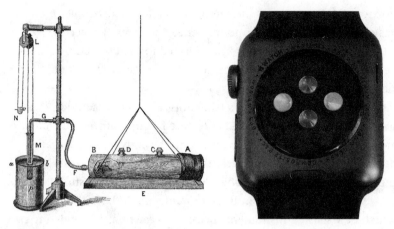

Figure 1.6
Left: Mosso plethysmograph (1876). *Right:* Underside of Apple Watch, with PPG (photoplethysmograph) sensor and LEDs. Photo by Fletcher.

In fact, not only was the graphic method imprecise, it was actually considered deceitful. Although there was a fervent belief in the all-powerful ability of these "self-registering" instruments—devices that automatically inscribed or self-recorded their data without human intervention—there was still a need for a human interpreter between the body and the instrument. The human eye had to read and understand the scales of data appearing on the drums of the kymographion or other machines.[51] Wrote one French researcher, "The registering apparatus does nothing but to inscribe undulating lines that fall on our senses; but once it comes to interpreting the traces, the graphic method has no more certitude than direct observation."[52]

Things now have changed. While the curves of Marey look suspiciously like those Fitbit app or Apple Health app curves on your smartphone, there are major differences. Direct observation replaced by automated algorithms and rows of network servers that store the statistical analysis of the world's sense data have become our new sensory physiologists. Now sensing machines capture, read, and analyze signals produced by human bodies—blood pressure, glucose, breathing, nerves—with even less recourse to human observers. Precise electronics, digital signal processing, statistical models, and the automation of computation are erroneously believed to have eliminated the imprecision of both the senses and mechanical instruments. If the data is imprecise, the algorithm can always be tweaked. The same goes for such technologies when they are revealed to actually exhibit cultural, gender, or racial bias. The technological solution is to identify the problem and quickly fix it, rather than recognizing the fundamental flaw in design assumptions in the first place.[53]

The experimental researchers of the end of the nineteenth century who sat with pen, paper, and instrument, ready to measure reaction to stimuli, have instead moved away from the scientific lab where they originally sought new knowledge about the human senses through sensors and into the reality labs of Facebook or Apple's secret "exercise lab."

Apple's lab would most likely be the envy of Wundt and his younger disciples. This secret facility in a bland Cupertino, California, building employs not only thirteen exercise physiologists and twenty-nine nurses and medics but also an army of machines to log tens of thousands of hours of subjects' physiological data as benchmarks to test the sensor-embedded products of the world's most valuable corporation. Apple is proud of its instruments. According to the lab's director, it has "collected more data on activity and exercise than any other human performance study in history."[54]

Comparing the emergence of sensing machines in the nineteenth century with today therefore reveals both a historical continuity and a radical break. In the 2020s, every individual who dons a fitness tracker, smart watch, biometric shirt, or wearable sensor engages in a process of transformation: turning oneself into a self-monitoring test subject without the intervention of the human psychologist or physiologist.

The myriad of sensing-measuring gadgets we now take for granted were still laboratory-bound in the nineteenth century. Instruments that logged physiological signals didn't leave the sites of experimental science for the gym or the office as they do now; they were instead parts of larger scientific apparatuses.[55]

Moreover, there is a fundamentally different understanding of the human in relationship to our technologies of digitization. In fact, even if human bodies then were at the whim of instruments that abstracted their senses into graphically plotted signals with nineteenth-century tools, there was still a connection between the person that produced the data and the resulting numbers. One could glance at the squiggly marks on the soot-covered surface of a drum after an experiment and claim, "that is me."

But the computational automation of mathematics and statistics has changed this. The way we understand the temporal role of sensing now is radically different. With the kymographion and sphygmograph (blood pressure) instruments, time was recorded graphically at different scales on the physical surface of a rotating drum or on a paper surface (figure 1.7, top), by manually speeding up or slowing down the mechanical instrument. The visualized curves generated by Fitbits and Apple Watches are different. They are the by-products of statistical processes: the size of a window through

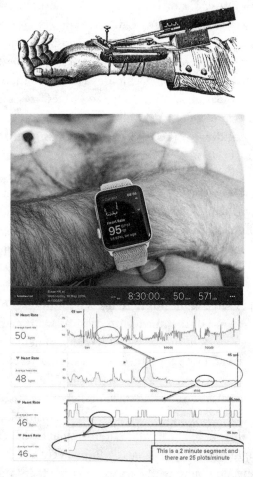

Figure 1.7
Top: Marey sphygmograph (circa 1885). *Middle:* Self-monitoring. *Bottom:* Different heart rate signals on a smartphone dashboard.

which you see only part of a longer and continuous signal,[56] or derived from statistical techniques (figure 1.7, bottom). In other words, the curves that are output represent an already computationally processed artificial time.

Like the big data world they are part of, in which meaning is dependent on the right mathematics to find patterns and meaning in a sea of randomness, our new psychophysicians and physiologists also believe that the truth of the senses can be found in the numbers—in statistical techniques that measure and predict the future based on the past.[57]

At first glance, when you click on the Fitbit dashboard, you can believe that world of colorful data represents the whole of you. The glowing curves on the smartphone or computer monitor seem to display *your* benchmarks; they reveal the shape of *your* heartbeat; they visualize *your* overall "performance" doing mundane workaday things like walking to the bus. But what is really output is only a fraction of ourselves. "We" are both before and after what we see in the graph on the screen. We might thus start asking *when are we* versus *who are we*.

The sophistication of sensing devices to govern our senses has radically shifted as well. As we will see, the instruments of Wundt's laboratory would undergo their own revolution. Through sophisticated changes in electronics and computation, they eventually would become attached to and interdependent with the bodies they would be measuring and, ironically, at the same time, divorced completely from them (figure 1.7, middle). And yet, this is not all. The technologies of sensing developed in the nineteenth century have a similarity to those sensors we wear on our bodies today: they also ignored the fact that bodies are different. The possibilities of the wearable sensor revolution are endless, but they both recall and radically revise and still render invisible those human subjects who produce new sensory knowledge, whether in the psychology laboratories of the nineteenth century or the cubicles of Facebook research in 2022.

Perhaps most importantly, the context and purpose of sensory measurement itself has radically transformed since Fechner announced his psychophysics. The physiologists and psychophysicists of the past, who turned to instruments to technologize themselves and their test subjects, now go an extra step. They now automate the 172-year-old scientific technique called *psychophysics* to *design* the next generation of perception machines. With today's sensing machines in our clothing, cars, houses, games, stores, theaters, and galleries, measurement thus goes hand in hand with design and creation. The need to probe the human senses with instruments and machines is not only about gaining knowledge about how these senses work; it is applying that knowledge to designing and perfecting systems that produce and anticipate new connections between our perception and those instruments and machines, where both expand each other.

In contrast to that which came before, our new sensing machines more accurately capture and analyze the microtime and microspace of our breath, heartbeat, brainwaves, muscle tension, or reaction times. But they do this for another reason. Our sensing machines now conceive and create techniques that aim to fulfill that long sought-after dream of those forgotten nineteenth-century researchers like Fechner and Marey: to become one with what Fechner called the *animated substance* of the technological world itself.

Playing

2 Accelerate!

> Effortlessness is one of the cardinal virtues in the mythology of the computer . . . Though the principle of effortlessness may guide good word processor design, it has no comparable utility in the design of a musical instrument. In designing a new instrument, it might be just as interesting to make playing it as difficult as possible. Physical effort is after all a characteristic of the playing of all musical instruments.
>
> —Joel Ryan, "Effort and Expression"[1]

In January 1995, I caught a firsthand glimpse of sensing machines. The location was unexpected: a rehearsal room at the Frankfurt Ballet, the ballet troupe of the city of Frankfurt, Germany. Directed by the American choreographer William Forsythe, known for his large-scale dance-theater works that turned the movement vocabulary of classical ballet on its head while incorporating the latest technological and theatrical explorations, the Frankfurt Ballet's radical approach to ballet made the company the envy of the international dance and theater world in the 1990s to the 2000s.[2]

One day, in the midst of a rehearsal for a new evening-length premiere I was working on as a production assistant, a sensor attached to a small box mysteriously appeared in the rehearsal room. This sensor would transform how the other team members and I would think about the expressive potential of a machine.

The title of the three-act contemporary dance-ballet-theater-multimedia piece we were creating was just as mysterious as the sensor: *Eidos:Telos*, two Greek words brought together that loosely interpreted could translate as "image: end" or "form: end." The title was not chosen by chance. It was inspired by passages from *The Marriage of Cadmus and Harmony*, a work of historical fiction by the late Italian writer, translator, and publisher Roberto Calasso that we were mining as a reference for our work. Calasso has long written genre-defying books on how ancient myths—Greek, Indian, African—provide alternative frames of reference and a modern understanding of such large subjects as the nature of ritual, the formation of Western consciousness, and the linkages between mythology and history.

During the preparation for the production, one of Calasso's stories—the Greek myth of Kore, better known as Persephone—became the basis for the ballet's second act. The Persephone myth warrants a brief retelling. Hades, lord of the underworld, carried off Persephone, the daughter of Zeus and Demeter (the goddess of the harvest), one day while Persephone was picking flowers.

Calasso describes this horrific scene in exquisite, poetic detail—a scene dealing with the forces of the invisible. "Where dogs would lose their quarry's trail, so violent was the scent of the flowers. And here, near Henna. Kore was carried off. When the earth split open and Hades' chariot appeared, drawn by four horses abreast, Kore was looking at a narcissus. She was looking at the act of looking. She was about to pick it. And, at that very moment, she was herself plucked away by the invisible toward the invisible."[3] As part of a deal brokered between Hades and Zeus, Persephone would split her time between the world of the dead ruled over by Hades and the world of the living on Earth. While Persephone spent six months a year in the underworld (creating winter on Earth), she returned to Earth from the realm of the dead, working with her mother Demeter to renew the fields and life itself or what we call spring.

This Persephone scene would eventually be a key dramatic element in *Eidos:Telos*. This was no vanilla Swan Lake ballet production. It would be a dance theater work "about seeing the end of things" and a "rage against death," as a New York dance critic later described the work when it toured in the US. These were appropriate themes considering that the Frankfurt Ballet was in mourning while creating this work after the death of Tracy-Kai Maier, one of the company's young star dancers and Forsythe's wife.

But at this early moment in a very compressed one-month rehearsal period, the music composition team I was collaborating with was not dealing with such grand metaphysical questions but with something far more mundane. We were working on the music score, which was to be performed live by a violinist and three trombone players. The score would not just be acoustic; it was to be manipulated live with electronics.

The music was the work of two composers: Forsythe's usual collaborator, Dutch composer Thom Willems; and his American colleague, a musician and scientist named Joel Ryan who is an expert at digitally modulating the sound of live instruments through an approach called *live signal processing*—transforming the sound of acoustic instruments into digital signals and then manipulating them into wild, electronic tapestries of sound.

At this point, however, the only music that had been written down was Willems's earlier score to be used in the first part of the ballet, a deconstructed adaptation of composer Igor Stravinsky's music for the 1927 ballet *Apollon musagète*. The core music would emerge from improvisations conducted by processing the sounds that the violinist, a Russian named Maxim Franke, made on stage as part of the musical score. *Processing* here means amplifying the violin with a microphone and then taking the resulting sound and shaping it by means of a rack full of electronic devices and computer software to get unheard of, hallucinatory sounds that suggested another world.

Drawing the bow hard against the backside of the violin, for example, created an overwhelming wall of noise. A long, sustained pitch produced a massive chord of celestial-sounding harmonies. As we experimented with various playing techniques, trying to get the violin not to sound like a stringed instrument, we also tried out different software instruments that altered the sound and then quickly noted the results in detail in notebooks.

During a rehearsal break, Joel Ryan unpacked an object wound in bubble wrap and what appeared to be laminated plastic. He revealed a small electronic component and attached it with a long, thick cable to another book-sized machine stamped with the label "STEIM SensorLab Controller Box." The component intrigued the team. Ryan then explained that this mysterious thing was an *accelerometer*—a relatively simple sensor that measures the change in velocity, or the acceleration, of an object as it moves over time.

Ryan brought the accelerometer along because he thought it might be interesting to use in the violent moment in the second act of *Eidos:Telos* when Hades snatches Persephone away from Earth. Dana Caspersen, the virtuoso dancer/performer who was playing the role of Persephone, was working with us to come up with the right mix of sound, movement and text. Caspersen improvised with different voice tonalities while the Russian violinist interrupted and punctuated her monologue with violent squeaks, plucks, and bowing on the strings. We processed the violin's sounds to make them ever more ethereal, almost otherworldly.

Ryan had an idea: attach the accelerometer to Caspersen and capture her movements with the sensor to control the intensity, pitch, and timbre, or color, of the music, together with the ceiling lights in the Frankfurt Ballet opera house. This would create the crucial dramatic moment when the ground opens up around Persephone and Hades kidnaps her from this world.

Both the accelerometer and the controller box came from the Studio for Electro-Instrumental Music (STEIM), an experimental cultural center in Amsterdam founded in the late 1960s and specializing in creating new hardware and software technology for musicians and where Ryan served as a scientific advisor.[4]

Ryan himself is a polymath. He knows a lot about everything in the sciences and the arts, but particularly how sensors, music, mathematics, and computers work together. After studying physics at Pomona College as an undergraduate in the 1960s, Ryan switched to philosophy at the University of California at San Diego with German countercultural hero Herbert Marcuse. In the summer of love of 1967, he also spent his spare time taking sitar lessons with Ravi Shankar, the world-famous Indian musician who famously taught George Harrison and welcomed the Beatles into the incredible world of Indian classical music.

Ryan also studied electronic and experimental music at Mills College in Oakland, California, supporting himself by working as a physics research associate at Lawrence UC Berkeley Labs, one of the world's premiere physics research centers, where he designed sensor arrays and real-time software for solar energy experiments. Back then, sensors were not separate components you could buy at electronics stores for a few dollars like today. They were embedded and expensive parts of big science measurement apparatuses used in multi-million-dollar research experiments.

At Mills College, a well-known women's college with a renowned, gender-mixed electronic music department, Ryan became part of a new generation of men and women musicians learning to program and use the first portable computers to compose, manipulate, and perform music in real time. *Real time* is a critical concept in the research area of *computer music*—programming computers to generate sound—and describes a condition by which musicians can modify the sounds a computer makes while the sounds are being output by the machine.[5]

As Ryan told me twenty-five years after we first worked together at the Frankfurt Ballet:

> Most of us in 2020 experience our idea of digital music making as instantaneous. You click and sounds appear. Real time seems redundant. But back in the 1970s to 1980s, that was actually not the experience or the expectation for most users of music software. By around 1978, there were just a handful of computer musicians who chose to play with digital machines in the musical moment, having to write their own software on small computers. They felt the need—I certainly did—to call this *real time*. But at the Lawrence Berkeley labs, there was already a real-time group where I had a day job! If a computer was to take part in an experiment, it was inconceivable the experiment would need to wait for the computer to catch up. So we designed computers that could run at the time scale of the experiment. Physics was about time itself, and real-time computing was necessary for experiments that were all about executing actions in time. Physicists had a need to get computers into their experimental practice.[6]

Before the 1980s, making music with computers was a thankless task. Computer music researchers—and occasionally musicians, in the rare moments they could get access to the machines—waited hours and sometimes days for the computers they programmed to make sound. Musicians could only produce sound by rendering it with expensive mainframe computers in institutions like Princeton/Columbia or Stanford. The experience we now have of waiting for video and high-end 3D images to render would hardly be the equivalent of the endless delays that musicians had to endure with the computers they laboriously learned to program. This temporal limitation of mainframe computers, as well as their size (they took up entire rooms) and cost, made their use impossible for live performance. In other words, real time was not only a technical achievement—it fulfilled the desire for musicians like Ryan to get musical results immediately so that computational time could be synchronized with lived, human time.[7] Ryan's (and others') belief was that musicians couldn't let computation control musical timing. Whatever the computer did, it had to follow the time of the music.

Real time, having the computer produce sound immediately, is also the reason that Ryan, a musician who is a world expert at building software systems to perform and manipulate sound in real time for artistic purposes, would have an accelerometer in the first place and bring it to a ballet rehearsal! Yet, Ryan uses his accelerometer in a decidedly unorthodox manner—as an *input device*, like a mouse or a keyboard, for his computer, using code he's written that can take the sensor's signal and use it to manipulate, shape, and control different factors or *parameters* of sound.

In the Frankfurt Ballet rehearsal room, Ryan demonstrates how the accelerometer works. He secures it to his hand with heavy tape and begins moving. The device comes alive, producing a flowing stream of numbers on the screen in software Ryan has written to capture the sensor's data. Moving his hand slowly and continuously produces little change. But a sudden jerk or a speeding up and slowing down yields a completely different response, at least in the numbers that seem to jump all over the place.

Ryan then tries another experiment. He connects the stream of accelerometer data to a sound—a simple, monotonous sine wave—and repeats what he did before. The sound blares out of the heavy-duty loudspeakers in the rehearsal room. When he continuously moves his hand in one plane, the sine wave just continues to play. When he does the same motion but starts to change the orientation of his hand, however, something interesting happens. The sound changes in pitch with the rotation of his hand on one axis (call it the y-axis) and changes its loudness in the other direction (the x-axis). A violent jerk causes the sound to stutter. A slow movement

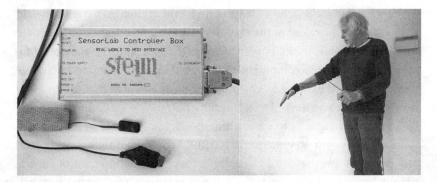

Figure 2.1
Left: Accelerometer with STEIM SensorLab. *Right:* Joel Ryan demonstrating accelerometer (2020). Photo courtesy Joel Ryan.

followed by a sudden speeding up distorts the sound in an unpleasant manner.

Ryan then tries to explain to his audience of nonphysicists what is going on, demonstrating with almost choreographed hand movements how acceleration works. Motion is all about change. You move from here to there. *Velocity* is the speed at which you moved to get there, but your speed can change too, from slow to fast. That change is what mathematicians name the *derivative* or acceleration of your speed in the branch of mathematics called calculus. An accelerometer measures this change as a stream of numbers: positive for speeding up, negative for slowing down, and zero when you move at a constant rate.

But what is our physical *experience* of acceleration? One is that classic example of stepping on the gas pedal, which causes a car to lurch forward and then take off. But slowing down by suddenly slamming on the brakes is acceleration as well. Moving continuously, like Ryan did with his hand at first, produces zero acceleration—no change.

Acceleration also happens in another somewhat counterintuitive manner—when you change the direction a vehicle is traveling in. Recall the experience of a roller coaster. The initial drop is hair-raising because the cars rapidly accelerate. You feel that you are falling, which is exactly what is happening as you experience the force of gravity. But the real sense of feeling the pull of the physical world is when you get acceleration from turning—on a roller coaster, when you enter a curve. The centripetal force almost yanks your bones out of your body, pushing you against the car and away from the center as you move through a hairpin curve. The same physical feeling happens when you exit a freeway going too fast and turn the steering wheel to navigate the curve in the road. Turning yields acceleration that has strong physical sensation; we have a built-in bodily sense of acceleration because we strongly feel its quantitative effects.

Inspired by Ryan's description and inventiveness, we spend the next days trying to work with the accelerometer, eager to experience how a dancer's movement, especially a dancer like Dana Caspersen whose body can twist and turn in space like a dervish, might also be converted into sound. This idea is ahead of its time. In other words, it doesn't really work. The cable gets entangled in Caspersen and she trips over it while attempting to whirl around. We tape the accelerometer to her hand but sweat builds up and the sensor loosens, giving less than ideal readings as its orientation is off.

The SensorLab, the control box needed to decode the electronic signal into meaningful numbers that Ryan can read with his software, is also fussy. It doesn't like it when Dana pulls too hard on the cable, disconnecting it from the box. It becomes difficult to reestablish communications between the sensor, the SensorLab, and the computer after this happens, making the whole system fragile.

Like many other ideas we have in the course of the rehearsals for this ballet, the idea of using the accelerometer in the production disappears after one week. It goes back into the bubble wrap and box. Yet in retrospect, it is telling why our team was so fascinated by the otherworldly potential suggested by this sensor. We glimpsed the creative possibilities of translating abrupt changes in velocity and direction from a moving human body into sound. We imagined how a dancer might generate physics that would shape ear-shattering sounds or sudden bursts of blinding light around her to express the apocalyptic moment in *Eidos:Telos* when Hades swallows innocence. The accelerometer attached to a dancer's body could form a perfect interaction of human and machine dancing together. The forces of the invisible had directly captivated us.

⌒

In 1995, you probably had never heard of an accelerometer, much less used one to make music. In 2022, this sensor has become ubiquitous in daily life. Each time you check your Fitbit for how many steps you took, measure how soundly you slept, have your hard drive rescued when your laptop falls to the floor, or readjust your phone to glimpse into a landscape portrait of an image, the tiny accelerometer exerts its presence. Perhaps more than any other sensor, the tiny accelerometer (figure 2.2) has had a critical impact on how we negotiate our digitally mediated interaction with the physical world that surrounds us: how we move, turn, walk, hold objects, play games, and create music.

It's not a surprise that the accelerometer was not widely known before its widespread use in smartphones, fitness trackers, and game controllers. The physics of acceleration are far from intuitive. The common understanding of acceleration is something speeding up. We accelerate when we step on the gas pedal in a car—a device that, in fact, is still called the *accelerator*.

But changing *speed*, the measure of the distance of a moving object over a given period of time, is only one aspect of acceleration. The direction in

Figure 2.2
Three-axis accelerometer in 2019. Photo courtesy SparkFun Electronics.

which something speeds up is also critical. Acceleration, in other words, is a *vector*—a mathematical quantity signifying both magnitude (in this case, **speed**) and direction. When an object accelerates, it can thus change **speed**, direction, or both quantities at the same time.

Acceleration is not just the change of velocity of a moving object. Things that vibrate or shake also exhibit acceleration, which is why early accelerometers were bulky and heavy mechanical devices designed to sense vibrations around and inside objects such as bridges and machines. Although earlier versions of accelerometers date back to the mid-eighteenth century, the modern accelerometer appears around 1923.[8] American geophysicists Burton McCollum and Orville Peters, working at the National Bureau of Standards, proposed an *electrical telemetry* device to measure vibration at a distance. *Telemetry* is the science of monitoring and collecting information remotely and then electronically sending it to another location.[9]

The fact that engineers were interested in measuring vibration at a distance also explains why accelerometers have long been used in geophysics to monitor the vibrations of the earth itself. Earthquake-detection devices called *accelerographs* embed accelerometers within them to precisely measure the *strong motion* of the ground during a powerful earthquake—ground motion of sufficient amplitude and duration to be potentially damaging to a building or other structure.[10] Accelerographs can monitor ground shock from an earthquake better as the power of such an event is easily outside of the measurement range of normal seismographs.

More modern industrial applications of accelerometers involve measuring *resonance*, the property that causes a mechanical system to begin to vibrate or oscillate more intensely at certain frequencies over others, in structures. Normally, such oscillations die down. If they continue for longer periods of time, energy builds up within the structure, causing it to vibrate ever more violently and, eventually, to break down. Inventor Nikolai Tesla's mythic and legendary earthquake machine, based on his 1894 patent for a reciprocating engine, aimed at demonstrating this principle.[11]

Tesla's unrealized invention inadvertently found a more contemporary context. Shortly after the celebrated opening of British architect Sir Norman Foster's Millennium Bridge in London in June 2000, the structure began wobbling violently under the footsteps of pedestrians who produced what engineers called *synchronized footfalls*—a sudden emergence of locked-in-step rhythms produced by thousands of people as they crossed the bridge at the same time.[12] This 320-meter-long concrete suspension bridge pile anchored into the Thames, which its builder, British architectural engineering firm Arup, called "an absolute statement of our capabilities at the beginning of the 21st century,"[13] became like a floppy but tensed rubber band, causing it to be shut down one day after its opening. Arup later ran real-world experiments before reopening the bridge, using video analysis and accelerometers to measure different levels of vibration produced by test groups walking across its surface.

If accelerometers were initially measuring vibrations in structures, they found new applications in the period after World War II. The introduction of jet propulsion in the mid-1940s gave the accelerometer a new military life and purpose. The US Navy began commissioning research into measuring different frequencies found in mechanical shock produced in fighter jets. The accelerometer was deployed in airplane cockpits to measure the force of gravity generated in takeoffs.

Accelerometers were also ultimately destined for space. They were used in the inertial guidance systems of intercontinental ballistic missiles (ICBMs) in the 1960s to measure the changing angle or tilt when the missiles were in air, as well as in Apollo 11 for vibration modeling, velocity measurements, and other data to locate the spacecraft's position traveling from Earth to the moon and back.

Accelerate!

Back on earth, however, Joel Ryan's laminated accelerometer that he had brought to the *Eidos:Telos* rehearsal room came from a more unlikely source. Around 1994, Ryan had heard that accelerometers were being used for a new purpose: controlling airbags in cars. Accelerometers sensed sudden changes of acceleration in cars and subsequently launched an airbag if the acceleration became too great.

Why use the sensors for airbags? Since Einstein, we now see gravity itself as acceleration and the most familiar experience of acceleration is our weight, which is said to be an acceleration of 1 g (gravity). As well as measuring changes in velocity, we can also use accelerometers as *orientation devices*, since the acceleration of gravity provides a static reference in one direction "down."

The number of g's indicates the amount of acceleration. When you are weightless, there is no acceleration—0 g. Standing on the ground is equivalent to 1 g. Humans can experience 5 g in a roller coaster. Fighter pilots are trained to handle sustained levels of 9 g acceleration. More surprisingly, everyday events also produce strong acceleration. Sneezing after opening a can of hot chili powder produces around 2.9 g while a hearty slap on the back is close to 4 g. Jet fighter pilots taking off and accelerating endure much more: 7–9 g, but only for a few seconds, and they are physically fit to withstand such pressure. Blackout and death occur when 5–6 g are sustained for longer periods of time.

The number of g's that can be produced in a car crash is more staggering: on the order of 40–70 or the equivalent of 2.4 tons ramming against a fleshy body! Here the accelerometer measures the continuous acceleration, anticipating a coming disaster and attempting to head it off. If the g-force of a moving vehicle reaches approximately 30–50 g, which takes place during the rapid deceleration of velocity when a car collides with something, the accelerometer triggers the air bag inflation mechanism electronically.[14]

The accelerometer could trigger an airbag when a vehicle changes speed, direction, or both at the same time. The airbag sensor is not just active when a car speeds up but if it rapidly changes velocity in any direction; if the car is hit from behind, causing it to lurch rapidly forward; the car brakes suddenly or is hit from the front; or the car is hit from the side, causing it to suddenly change orientation.

Like many enterprising artists who knew their way around technology, Ryan was intrigued when he discovered that the universal characteristics of motion were wrapped up in an electronic chip. He called up the small California corporation providing accelerometers for experimenters in the not ready for primetime airbag industry and asked for a sample, receiving one of the devices in the mail soon after—a smallish chip mounted on a white ceramic plate that could withstand real-world conditions. What excited Ryan was that a stream of numbers produced by a sensor could capture the immense complexity of our kinetic experience of the world; the curving grace and swerving, vibrating richness of motion (which marks all dance and music) were suddenly available to be experimented with.

How then did a sensor meant to measure acceleration for use in cars end up as an expressive device to make and perform music with computers, as Ryan was already doing in the mid-1990s?

At the same time that Ryan was discovering the expressive power of the accelerometer, a radical technological breakthrough took place with the development of so-called microelectromechanical systems (MEMS). Imagined in the 1950s, MEMS technologies follow the spirit (if not the letter) of Moore's law, Intel cofounder Gordon Moore's famous statement that the number of components on a microprocessor would double every eighteen months.

MEMS components imitate the electromechanical world at miniature scale. They involve the design and fabrication of complex mechanical systems like levers and springs together with wires and electrical components that all fit on a single chip. In fact, in a 1959 lecture entitled "There's Plenty of Room at the Bottom," legendary physicist Richard Feynman gave a taste of what was to come with MEMS technology. "Why cannot we write the entire 24 volumes of the *Encyclopedia Britannica* on the head of a pin?" Feynman provocatively asked. His solution was bold. The future of solid-state physics and electronics would be the ability to manipulate, manufacture, and control things at atomic and molecular scales.[15]

Feynman not only focused on how to use Bell Labs's engineer Claude Shannon's statistical conception of information as the number of binary digits or bits required to encode the messages that could be written on a pinhead. He also theorized how to manufacture the electronic and computational components to manipulate atoms and molecules similar to how we control levers and engines. "Why can't we manufacture these small computers somewhat like we manufacture big ones? Why can't we drill holes, cut things, solder things, stamp things out, mold different shapes all at an infinitesimal level?"[16]

It's no wonder a physicist like Feynman celebrated atoms and molecules. Both are measured at nanoscale: one billionth of a meter. His provocation would eventually become reality, one of the de facto inspirations for the subsequent development of nanotechnology, the manufacturing of components at nanometer (nm) scale. When they began to appear in the 1980s, MEMS sensors and actuators combined together on a chip were considerably larger—between one and one hundred micrometers (1,000 nm—100,000 nm). A water molecule is between 0.2 and one nanometer. A germ is approximately 1,000 nm. Even larger, a human hair could be measured at around 100,000 nm. This was still extremely tiny, considering that one nanometer is the equivalent of 0.000000001 of a meter.

Feynman's call for new fabrication techniques found its way into MEMS designs. The mechanical world of cantilevers, levers, and springs at a scale barely perceivable by a standard microscope appeared on chips that you could hold in your hand but whose components were so small it was impossible to see them with the naked eye.

The accelerometer was indeed one of the earliest nanotechnologies and became the perfection of MEMS thinking, with its industrial legacy reimagined for an age now dreaming in atomic and molecular scales. The cumbersome, mechanical monstrosities that had prevented the accelerometer from being used in everyday situations disappeared, replaced by wafer-thin sets of electronics: in one standard form, a tiny floating mass attached to a spring suspended inside an outer casing. Acceleration, a sudden shock, or a vibration would jostle this mass, causing the tiny spring to stretch with a force that would register the acceleration on one or more axes—x, y, and/or z.

The shift to MEMS made possible the deployment of the accelerometer in hard-to-reach places, like the bumpers of cars, the interiors of handheld remotes, or digital assistants. MEMS also made possible their use in a more unlikely application: a new genre of musical instruments as sensing machines, just like the kind that Joel Ryan had begun to use for music making.

One year before the *Eidos:Telos* rehearsals, Ryan had used the accelerometer he had brought for another artistic purpose: to control, shuffle, jump around, and "scratch" a video that was encoded onto video-based laser discs and played back with a laser disc player, like a DJ manipulating vinyl.[17] Here, a small set of intuitive gestures asserted a wide range of quantitative control. Rotating his hand back and forth on one axis controlled the precise speed and direction of the video playback, from high-speed scratching to standard play or freeze frame. A more vigorous jerk toward the body or a thrust away could jump the playback forward or back from a few seconds to a minute or two. Through his gestures and movements captured by the accelerometer, Ryan was trying to find a way to expressively use the sensor to "play" the laser disc like a new kind of musical instrument.

Despite its engineering and scientific origins, the accelerometer lent itself well to such expressive use. That the sensor could capture a wide range of changing intensities or forces of musical "gestures," from large and impulsive ones like striking a violin, jerking a hand across a keyboard, or beating a drum, to smaller, more nuanced kinds that involved low energy like tiny plucks on a string or even filigreed finger movements, all in real time, was one of the main reasons that musicians and researchers began exploring accelerometers as potential interfaces in their search for new technologies for musical expression. What couldn't be seen by an audience was still essential to musical expression.

Like Ryan, musicians sought to integrate accelerometers into acoustic instruments or even build their own new augmented instruments or *musical controllers*.[18] Such musical contexts cannot be overemphasized. They represent a momentous context change in how we think about sensors, where such technologies are no longer seen as simply passive measurement devices but as new instruments for artistic expression. This interest in what were soon dubbed *gesture-based musical interfaces* (see figure 2.3) exploded in prestigious computer music engineering laboratories at MIT and Stanford and at composer/conductor Pierre Boulez's government-supported, Paris computer music center IRCAM—but also at smaller organizations, like STEIM, the electronic music research studio at which Ryan worked and eventually, in the studios of independent artists. Enterprising, technology-savvy musicians and visual artists were quickly driven to explore the expressive potential of the accelerometer.

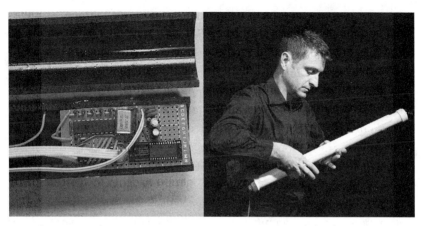

Figure 2.3
Left: Example of a gesture-based musical instrument called the "T-Stick" with accelerometer electronics inside. *Right:* Musician D. Andrew Stewart playing T-stick. Courtesy Joseph Malloch; photo by Vanessa Yaremchuk.

The pop music world also jumped on the bandwagon.[19] In the late 1980s, the Zeta music company, which built electronic guitars, created the Mirror 6: an "augmented" guitar that featured accelerometers to measure the degree of shaking produced while playing the instrument. In the late 1990s, engineer Bob Bielecki and researchers at Microsoft founder Paul Allen's Silicon Valley–based think tank Interval Research designed and built a wireless harpoon-like instrument using an array of sensors, including an accelerometer. The instrument was for musician Laurie Anderson for her 1999 music theater work *Songs and Stories from Moby Dick*.[20] Sporadically during several sections of the evening-long multimedia concert, Anderson appeared on stage, brandishing the long object, which produced strange kinds of sonorities that awed the audience when rotated and shaken.

But the accelerometer as an instrument of musical expression also had drawbacks. There is an inherent tension between the physically engaged act of playing a traditional musical instrument versus the mathematical abstraction of the computer processes that could capture, analyze, and connect data from the accelerometer to influence sound. Ryan carefully described this issue when he wrote, "The need for 'hands on' in performance forces a composer to confront the abstractness of the computer head on. Each link between performer and computer has to be invented before anything can be played."[21]

Using sensing machines to make music is thus no trivial matter. To master them requires technical skill, movement dexterity, and an imaginative sense of how movement, data, programming code, and sound can mutually influence each other. Sensors like accelerometers are not arbitrary devices. They do not connect to the computer in a plug-and-play manner; nor does music burst forth without understanding the data produced by the sensor. Figuring out how to connect the sensor to processes in the computer all the while trying to achieve a level of expressivity similar to the playing of an acoustic instrument takes years to master.

Composers and programmers must therefore create models in the form of *algorithms*, sets of instructions in software, that will give physical form to their musical ideas—what Ryan aptly terms *physical handles on phantom models*. In other words, the interface or the controller shapes what kind of algorithms will be written to produce sound, as well as what kind of sound we eventually can experience.[22]

The accelerometer is a strange device. As a sensor, it is both highly abstract and oddly concrete. On the one hand, it measures acceleration, an extremely general feature of physical systems. Vibration and acceleration can be produced by any physical object or body, from a roller coaster or movement of the earth to a dancer falling to the stage floor.

In contrast, the context of what other sensors can measure is much more limited in scope. A carbon dioxide sensor that monitors the air quality in a closed building calculates a highly specific combination of gas molecules. A photocell or light-dependent resistor captures another phenomenon: the amount of light that strikes it, which produces a change of resistance in the light-sensitive material that lines the top of the sensor. In other words, these sensors measure specific and quite narrow things.

But attach an accelerometer to something and it reveals the complex and messy nature of our physical world. The signals the sensor produces are noisy and continuous: a stream of ever-changing values. To stem this flood of data, we need different types of statistical methods to smooth, and extract useful information from the lively and vibrating world that pours into this abstract device.

The accelerometer thus collaborates with the physical world. With its history rooted in the geophysical sciences, industrial measurement, and jet propulsion, the sensor measures shock against a body due to extreme changes of gravity or the sudden impact of an explosion or other violent disruption.

What does this notion of shock then imply? Perhaps it suggests that bodies, whether an ignition fuse, a pipe, a bridge buckling, an airbag blowing up, or a human dancing or making musical gestures, are actually quite similar. The accelerometer doesn't differentiate the machine from the human body or prefer it and vice versa. Both produce similar physical results. Instead, it's the scale and the context that matters and gives what we call *meaning* to the data that the sensor captures.[23]

3 "You Are the Controller"

> Today, if you don't understand the controller, you're not able to enjoy video games.
> —Iwata Satoru, former Nintendo CEO

In late 2006, an unusual phenomenon took place in living rooms across the world. Enthusiastic computer game players who had recently purchased a just-released, Japanese-manufactured game console with a novel handheld controller became a little too enthusiastic. As they intensely engaged in game play with the computer graphics bursting across their screens, swinging a virtual bat or golf club, rolling a virtual bowling ball, or bouncing in imaginary judo sessions, the controller in their hands took on unintended ballistic qualities.

Slippery from sweaty palms and fingers, the device suddenly disengaged from the human players wielding it, ending up lodged in the shattered glass surfaces of expensive LCD screens, in the faces of brothers and sisters, or literally on the lawn, courtesy of smashed living room windows. To this day, the internet is still rife with videos of users losing control and turning their controllers into dangerous projectiles.

TVs, plate glass windows and living room lamps were not the only casualty of the game controller's runaway success. In the late 2000s, articles began appearing in professional medical journals describing bodily injuries directly related to the use of the gadget, ranging from the standard (hand lacerations, bruises and tendon injury) to the more unusual, such as clavicular fracture (breaking the collarbone).[1]

Nintendo, the Japanese game company that invented this controller, known as the Wii Remote or Wiimote, was beset by a flood of lawsuits, eventually issuing warnings on its website to "allow adequate room around you during game play and to stay at least three (3) feet away from the television."[2] All the while, the company rushed to replace the Wiimotes' constantly broken security straps with newly "enhanced" ones.

Broken straps designed to keep the Wiimote safely tethered to the player's hands weren't the only reason the device ended up embedded in the glass front of televisions around the world. The deeper reason, and that of the Wii's success, lay in something far smaller: the 4 mm × 4 mm × 1.45 mm accelerometer embedded into the remote.[3]

The Wiimote was an overnight sensation worldwide when it debuted in North America on November 19, 2006. Affording a wide demographic of computer game players a new level of physical engagement in what had formerly been an industry of joystick and paddle interfaces that exercised merely thumbs and hands, Nintendo cashed in. Between 2006 and 2013, when it was discontinued, Nintendo sold over one hundred million Wii consoles at $249.99 a pop.[4]

The Wii's success story seems to be that of the first commercially successful game console and controller to enable a new kind of player experience by sensing player action and then decoding it into machine commands. But a 2010 *New York Times* article entitled "Motion, Sensitive" hinted at the deeper cultural impact of the larger trend of sensing human motion to power games.

A new generation of game controllers like Power Gig (an improvement on the wildly successful Guitar Hero) or Rock Band 3 had successfully bridged "the gap between media and actual person experience."[5] The Wii was one of these new controllers. Seen as a hallmark device, it represented a new reality in which a player's body became both game controller and interface simultaneously by directly translating motion from the physical world into screen action.

The results of this shift would not only advance the gaming industry but also have a far more wide-reaching impact. This was an effect that had been long sought after not only by international video game titans but also by another more unlikely group: avant-garde and experimental artists, poets, musicians, and thinkers who, since the early days of twentieth-century modernism, had sought to remove the gap between life and art.[6]

"You Are the Controller" 63

Figure 3.1
Wii controller in play.

How could sensors designed for games remove the divide between everyday life and art? To answer this question, we need to engage in some archaeological detective work to understand why sensors ended up in home gaming controllers in the first place.

Although the accelerometer is ideal for capturing the rapid changes of human motion that the Wii console video games demand, it wasn't the first sensor to transform games into human motion-sensing machines. Sensors in game controllers that could capture human motion date back as early as the 1980s. From the Zapper light gun outfitted with light sensors (for the famous *Duck Hunt* game from 1984) and pressure sensors in punching bags and dance-fitness pads to touch sensors in toy bongos, sensing technologies were already developed for the Nintendo Entertainment System (NES) and other game platforms.

These sensor-activated visions were test runs for the future of gaming. In 1993, Sega, another Japanese video game giant, developed one of the earliest motion-control sensor platforms, called the Activator—a system that appeared to foreshadow the Wii by eleven years but achieved little of its success. The sensor-based platform was designed for the Sega Genesis console system in which a player would stand inside an octagonal ring wired with infrared emitters that shot invisible light beams toward the ceiling.

The Activator was based on the concept of a light harp created by Israeli inventor and kung fu expert Asaf Gurner, who eventually patented what he called an "optical instrument, comprising tone signal-generating means comprising emitter and sensor means and means for producing tone signals responsive to signals produced or transmitted by the sensor means."[7] The Activator technology functioned in a similar way. Working on the principle of infrared sensing, when a player interrupted the invisible IR light beams, the system would supposedly register this break and use it to drive game parameters.

An amazing 1993 video of famed MTV video jockey (VJ) Alan Hunter describing the Activator at the Consumer Electronics Show (CES) in Las Vegas sounds more like a premonition than a sales pitch:

> Interactivity is the go word for any kind of new technology, right? Interactive video. Interactive multimedia stuff. But today we want to give you an incredible

Figure 3.2
Sega Activator box (circa 1983).

> sneak peak of something that will give new meaning to the word interactive. The one thing that's missing from video games, for teenagers I hear this a lot, is that they don't get to move their body enough. By god, when you've got that teen spirit, you want to move your body a lot when you're playing a video game. And this is what we are going to show you today. Something that is very physical and interactive.[8]

Despite heavy promotional videos demonstrating kids making fluid karate moves, the system was essentially a keypad with each of the eight IR beams functioning like a button on a standard game controller. The sensors didn't open up the expressive power of human motion. They constrained it, rendering players' bodies as glorified light switches.

Activator was the state of the art in another, albeit unintended, way. Like so many technologies before and after it, manufacturer desire outpaced technological capability. While Activator ads suggested plug and play, computer industry speak for a device that works perfectly with a computer the minute it's connected, the instructions to get the system to actually function demonstrated something else. The platform was described as inherently difficult to set up—an appropriate assessment considering that players (and, most likely, parents on Christmas day) had to engage in a step-by-step routine essential for many sensors called *calibration*—which, while well known to engineers, is unknown to consumers.

Calibration describes the process of adjustments performed on a sensor so that it can respond as accurately as possible to the thing it is designed to measure. Sensors are industrially manufactured electronic technologies, always subject to inaccuracies in measurement caused by component failure, changing environmental conditions, or manufacturing errors. Every sensor has what is called a *reference standard*. Think of this as the ideal measurement without any error. The concept of *accuracy* thus establishes a relationship between the reference standard and any new measurements performed by the sensor. The smaller the error between the two, the more accurate the sensor is said to be.

We assume that technology, and especially electronic technology, is magical and foolproof—until it fails. The calibration routine necessary to get the Activator to work must have struck its users as downright obtuse at the time. A long set of instructions declared the following: "Very Important. The Activator must adjust itself for optimum performance to the specific game play area. Stand Three (3) Feet away from the Activator and then turn on your Genesis. Wait for about 20 seconds or until the Game Title appears on the Screen. Remember to follow these instructions or your Activator will not work." The fact that a player would have to recalibrate the platform every time it was turned on or off or when a new game cartridge was introduced led to scores of frustrated users.[9] The instructions seemed to almost laugh in the face of the players when they claimed, "The system seems simple, because it is."[10]

While technologically and experientially the device left much to be desired, Sega's advertising was well ahead of the curve. Its tag lines and slogans all appeared designed to bring players' bodies into the loop, in sometimes almost violent ways. One early advertisement boldly claimed: "SOME KIDS WON'T SEE THE ADVANTAGEs OF ACTIVATOR. THEN IT WILL HIT THEM."[11] Failure notwithstanding, the Activator's tag line in its advertising promo foretold what was to come: *You Are the Controller.*

⌒

Four years earlier, another vision of "you are the controller" had also arrived. In 1989, the mammoth American game company Mattel released a much-ballyhooed sensor-loaded gaming device that was said to be lightyears ahead of its time. The Mattel Power Glove was futuristic in marketing speak, but toy-like in appearance. It resembled a thick plastic-looking glove and was designed to steer the Nintendo Entertainment System via hand gestures. It referenced a not yet tomorrow, when the advertising claimed that by putting on a Power Glove, "you put on the power of the future."

AGE, a New York-based entertainment company, saw the economic potential of newly emerging computer interfaces, what were then called *peripherals*, to give users a more "embodied" approach to computing. The sense of you having a body was a central feature of the game. The overall experience of interacting with a computing machine was about to be radically recast.

In 1981, the so-called personal computer had met its coming destiny. Human-to-machine interaction exploded with the commercial introduction of the mouse, ushering in a new era of computer peripherals transcending text-based interfaces. An even more exotic peripheral in particular struck the AGE engineers' eyes as ripe for a wider game-playing public—the Data Glove. This $10,000 sensor-riddled device was created by an early

Silicon Valley virtual reality company called VPL, cofounded by dreadlocked computer engineer and musician Jaron Lanier.[12]

VPL lives in the lore of Silicon Valley legend. Lanier is said to have coined the term *virtual reality*, despite the fact that the earliest use of the expression can be traced back most likely to the radical, early twentieth-century French playwright and actor Antonin Artaud. In a famous 1932 essay entitled "Theater and Alchemy," Artaud argued that the theater is a "stand in," a pretend-like doubling of reality that he dubbed *realité virtuel*. "All true alchemists know that the alchemical symbol is a mirage as the theater is a mirage," wrote Artaud in reference to the proto chemistry practice of seeking to turn base metals into gold. "This perpetual allusion to the materials and the principle of the theater found in almost all alchemical books should be understood as the expression of an identity (of which alchemists are extremely aware) existing between the world in which the characters, objects, images, and in a general way all that constitutes the virtual reality of the theater develops, and the purely fictitious and illusory world in which the symbols of alchemy are evolved."[13]

The theater, like alchemy, would be only a simulation. In 1986, however, the Data Glove *was* alchemy. Invented by VPL engineers Thomas Zimmerman and Mitch Altman, the cyberpunkesque apparatus with a glove made of neoprene could sense finger flexion (bending) and joint angles. Expensive fiber optics ran over and through the fingers of the glove. One end of the fiber optic featured an LED (the light source) and the other end a photoelectric sensor that could measure the amount of light from the LED interacting with it. When the user bent a finger, the break in the light beam caused by the bend in the fiber optic cable was captured by the photosensor, converted into an electrical signal, and read by a tethered computer that calculated how much the finger was bent.[14]

The glove could also determine where the wearer's hand was in space by using a high-end magnetic-based tracking sensor made by Polhemus, a company headquartered in rural Vermont, far from the innovation-driven frenzy of Silicon Valley. The $3,000 sensor in the glove was also the state of the art, from somewhat auspicious origins.[15] Originally developed for tracking the head position of military pilots for weapon control, Polhemus's sensor sent out electromagnetic signals via radio frequency to a receiver. Based on the signal strength measured between the sender and the transmitter, nine different measurements could be used to determine the position and orientation of the receiver—that is, the hand.

The Polhemus system featured a major innovation for tracking human motion—six degrees of freedom measurement—measuring forward, backward, up and down, left and right, and rotation around the three x, y, and z axes. The sensor could thus locate a moving object, like a hand, more precisely in space—one of the main reasons this sensing technology now proliferates in the commercial head-mounted displays used in VR and AR.

VPL's Data Glove bypassed the family recreation room, ending up at hallowed research institutions tied to the US defense complex, such as NASA and MIT, as well as in telemedicine and robotics. But AGE's founders also saw the increased market potential in the new ways we could communicate with machines—something called *natural interaction*, in which the computer would effortlessly disappear into the background of use.

The concept of natural interaction was a driving force in the development of peripherals and interfaces that would involve user's bodies. In fact, this notion emerged down the road from VPL's Redwood City, California, headquarters in another Silicon Valley incubator of the future: the research laboratory of the Xerox copier corporation, called Xerox PARC (Palo Alto Research Center, now simply known as PARC). While inventing the computer technologies we take for granted today in its research labs—the GUI, the mouse, Ethernet, and the laser printer—Xerox PARC was also laying the groundwork for a new model of interacting with computing machines.

One of Xerox's engineers, the late computer scientist Mark Weiser, was working on the "office of the future." Weiser's vision went beyond wanting

Figure 3.3
Left: Mattel Power Glove, 1989. Photo by Evan-Amos. *Right:* VPL Research Eyephone head-mounted display and Data Glove, with Polhemus tracking system.

to make computer peripherals more embodied. He sought to make them disappear altogether. Weiser called his concept *ubiquitous computing*, suggesting that computers would increasingly become invisible in their use.[16] They would move from our direct attention to what he called the *background* of everyday life. "There is more information available at our fingertips during a walk in the woods than in any computer system, yet people find a walk among trees relaxing and computers frustrating," Weiser articulated in a popular *Scientific American* magazine article in 1991 that introduced his research to a broader public than just computer scientists and engineers. "Machines that fit the human environment, instead of forcing humans to enter theirs, will make using a computer as refreshing as taking a walk in the woods."[17]

This naturalization of computing, of taking the natural "human environment" into account, suggested a powerful idea: that interaction with machines is not only through writing or language but also through our bodies. What psychologists call *embodied* or *sensorimotor* ways of interacting with the world that involve the sensory and motor parts of the body are second nature. We don't think for a moment about how we walk, that is if we are nondisabled. We just do it. Even those with reduced motion still engage in expressively interacting with the world, using their eyes, fingers, or other parts of their bodies. Likewise, when we dance, we begin to move our bodies to specific rhythms, not making conscious decisions about why we might thrust our legs out at one point or jump up and down at another. Some engineers felt that computing should follow suit.[18]

Ubiquitous computing's natural interaction concept certainly influenced the design of the Data Glove, as it would future visions of gaming. Users of the Data Glove would also find interaction with this strange new peripheral device "natural": they would forget that it was part of an artificial machine and come to recognize it as a part, even an extension, of their bodies.

In a bit of historical irony, the Data Glove was pitched at the newly emerging wave of virtual reality industries in the same period. At the time, VR marketed a future of disembodied simulation, a world without the pesky mortal coil of fleshy bodies. Yet bodies were somehow always lurking in the shadows. The interfaces to control virtual simulations were deeply material and tied to the physical world: weighty helmets, backpacks with computers, and the natural interaction imagined with the neoprene Data Glove worn on the hand.

But AGE's acquiring of the Data Glove license to produce the Power Glove with Mattel required radical changes to the glove's high-end sensors. Its fiber optics were replaced with cheaper and less reliable conductive and resistive ink as a new kind of optical flex sensor, while the prohibitively expensive Polhemus trackers would be substituted with something more robust that could withstand the shock and abuse that tween game players would inevitably dish out.

The engineers replaced the costly Polhemus system with cheaper ultrasound sensors. Similar to the echolocation scheme used by bats, ultrasound sensors measure distance based on the sending and receiving of ultrasonic waves—frequencies that are beyond the human hearing range. Two ultrasonic emitters on the glove sent a series of high-frequency ultrasonic beeps to receivers or microphones listening for the beeps and housed in a bar-shaped object mounted above the game display—the standard TV screen. The system used a process derived from surveying geometry called *triangulation*, which locates an object by taking measurements of it from two remote points. Based upon the difference in time between the transmitters sending the beeps and the receivers detecting them, the Power Glove could miraculously locate itself in space.[19]

Despite its toy status, the Power Glove promised an entirely new universe of natural and intuitive human-machine interaction for playing games through sensing. "The glove knows how you work," proclaimed one of AGE's partners and engineers. "You just act naturally and it automatically translates that into commands."[20] With technology industry hubris, Mattel, which bought the distribution license from AGE, imagined stratospheric sales. One million Power Gloves would be sold in the first year, it thought. But the seventy-five-dollar (or $155, in 2020 terms) Power Glove spectacularly failed. It sold only one hundred thousand units and could only be used to control two games. Its lifespan was also spectacularly short: one year.

Despite failure, however, something stuck. The motion control possibilities envisioned by the Power Glove would haunt the imaginations of game development behemoths' R&D teams for years to come. The Power Glove's quasi-spiritual-sounding claim to make "you and the game one" would be furiously carried forth in marketing speak and engineering quests. Indeed, it was the decision of Nintendo's president Iwata Satoru in 2002 to advance the ever-intensifying game console "arms race" with a new playing paradigm, the code name of which was Revolution—an appropriate moniker for a gaming system aimed at building a new customer base far beyond the standard teenage video game addict.

The future of Revolution (later called the Wii) was not just expanding Nintendo's customer base. It would also launch a new kind of technological imaginary using sensors that would advance this oneness of player and game to the next stage. The Wii catalyzed a whole new level of sensing machine within the context of play. It would facilitate a novel kind of experience by using a new sensor (the accelerometer) to help radically rethink the relationship between the gamer and the game itself and, in the process, retool how we understood interaction. In a series of interviews after the Wii's launch, Takeda Genyo, Nintendo's R&D general manager, explained that a game controller, the literal interaction device between the human player and the computing system, was not just a mouse substitute but could be something far more profound: "an intermediary between man and machine, and even an extension of the human body."[21]

Others had failed, so how would Nintendo realistically align the dream with the limitations of technology? By partially relying on what it already knew. Contrary to the idea that the use of the accelerometer in the Wii was a never-before-tested technology, the company had been quietly building up experience with it. Nintendo's approach was what economists call *path-dependent*—in which present and future innovation is driven by the past.[22] Not only does previous history matter; technological innovation and change also might appear to be more innovative than they actually are. Innovation occurs not in a planned way but because of random events or specific local conditions, such as a fluke or a one-time success with something that then gets ingrained as the only alternative. It might also occur because of previous experience developed by using a specific technology or a process.

As the early attempts at motion control using players' bodies demonstrate, the path to players thrusting their Wiimotes through living room windows was already shaped by previous attempts—observing both competitor's failures and Nintendo's own success in reconfiguring games into new kinds of sensing machines.

Ikeda Akio, a Nintendo research engineer, understood the potential of newer and smaller sensor technologies to transform the player experience.

In the early 2000s, he had deployed accelerometers in two Nintendo games: *Yoshi no Banyu Inryoku*, released in Japan, and *Kirby Tilt 'n' Tumble*, both designed for the earlier handheld Game Boy Color. The games embedded accelerometers directly into the game cartridge itself. Because the Game Boy Color was handheld, players wouldn't have to think about how to engage the sensor that registered how much the player tilted the device in order to move characters on the screen.

The manufacturing of Nintendo's mass-market accelerometer fell also to a path-dependent choice of company: Analog Devices, an East Coast US manufacturer of integrated chips and analog signal processing equipment and the key inventor of the MEMS accelerometers that had gained fame in the airbag-monitoring industry a decade before.

The choice of Analog Devices was not by chance. The company had supplied an earlier two-axis (x and y) chip that drove the *Kirby Tilt 'n' Tumble* game. In its early press releases, Analog Devices wasted no time announcing how its partnership with Nintendo would create "a true to life gaming experience." It also demonstrated that the firm had the engineering expertise from other fields to deploy in developing the game, perhaps to assure potential consumers of its technical abilities. "Analog Devices offers unparalleled experience with integrating motion sensing to enhance the products we use every day," one press release announced, "whether it's the automobiles we drive, the mobile phones we use, or the games we play."[23]

The Wii's breakthrough is by now well known—if not a bit surprising, even to Nintendo itself. Its commercial success was so swift that within six months, one of the company's main rivals, the Japanese electronics colossus Sony, also introduced an accelerometer into its SIXAXIS controller for the PlayStation 3, followed by another motion-sensitive system, the PlayStation Move, which featured a camera that could track user motions at thirty frames per second.

In the background, however, Nintendo knew that another "revolution" was brewing. Four years after the Wii's debut, the leviathan Microsoft upped the sensor-player ante with its Kinect system. Kinect was the apotheosis of a sensing machine for its time. Endlessly hacked and reengineered by artists and amateur technologists, it combined a color camera and a specialized sensor called a complementary metal-oxide-semiconductor (CMOS).

The sensor could "see" in 3D by detecting the distance between bodies and objects in collaboration with *machine learning*–driven software that identified different body positions and orientations in space—what computer scientists call *poses*. Indeed, the millions of consumers who bought and used the Kinect with Microsoft's Xbox 360 console in 2010 were probably not aware that their bodies were already being recognized and tracked using artificial intelligence.

For the Kinect, your own body in front of the camera was not enough. To recognize different poses, Microsoft had to collect the body data of others in order to train its algorithms to "see." The company's deep pockets facilitated camera crews sent to ten households around the world to record different body sizes and shapes moving and playing in their living rooms in endless combinations for a physical game that did not yet exist. Clusters of networked computers then processed these images, trying to predict the difference between an arm, an elbow, and a head—different parts of a player's body—and then storing this massive amount of data to later be recalled in the heat of game play.

In 2010, the Kinect's technological machinery was an astounding feat. Microsoft marshalled a who's who of just-out-of-the-lab techniques to advance the body-gaming paradigm: infrared light projection, CMOS sensors, depth detection, facial recognition, body tracking, edge detection, object distance measurement, and object recognition, together in one 66 x 249 x 67 mm set-top box.

For Microsoft, motion sensing like the Wii used was déjà vu; the company itself had already been there. Through a 1988 collaboration with Logitech, a maker of computer peripherals, it had dipped its feet into sensor-based controllers, developing the SideWinder Freestyle Pro and the WingMan Gamepad Extreme, both equipped with XY-tilt motion sensors, the same Analog Devices ADXL202 two axis accelerometers used in Nintendo's early cartridges.

Whereas the Wii demanded holding the twenty-first-century equivalent of a 1950s living room remote control, the Kinect sensor required no such device. It abandoned all attempts at approaching control through objects the player would hold in their hands. Distinctions between body and background, game and reality would fall away. There would be no separation between the room the player was in and the world of the game. The game was the space directly in front of the Kinect's cameras and nothing outside of it. The "you are the controller" mantra that had graced failed sensor controllers in the past, like the Activator, was resurrected from the ashes to finally reach its epiphany.

Microsoft wasted no time in sharing these epiphanies with its investors, press, and public. As the Kinect launched, Steve Ballmer, then Microsoft's CEO, declared in a propitious announcement that "we have entered a new era of 'natural user interfaces' where human-machine interaction would become as 'natural' as "touch, speech, gestures, handwriting, and vision."[24] Natural interaction once again reared its head. The player's movement in the game *was* the interface and experience rolled into one. The assumption was that you start moving and the machine does the rest. It was, as designers of interactive systems like to claim, "seamless."

But human-computer interaction is not particularly "natural." Gestures, movement, and action to play a game are inevitably culturally learned, just as Ballmer's list of natural interfaces like speech, gesture, writing, seeing, or hearing are also learned. Gestures hold symbolic weight, expressing cultural and social assumptions. Language is not just given in the head but shaped by one's social context. Seeing is learned through repetition, movement in the environment, and the concrete situation and its physical dependencies.

The same culturally learned rules apply to human-machine interaction. Programming a machine is learning the syntax and semantics of an artificial language. Moving a cursor is not something one is born knowing. Dancing before an optical sensor that constrains movement to a specific

Figure 3.4
Playing in front of the Microsoft Kinect sensor.

area of a room based on its field of view is different from walking through and sensing the forest, despite the fact that both activities require locomotion, gait, and sensorimotor action.

⌒

A more important challenge to the concept of "natural interaction" can be found through a closer look into the first generation of the Kinect's sensor and software architecture.[25] At first, these details could be thought of as purely technical. But they reveal something deeper, not only about the scale and sophistication of Microsoft's technology in the Kinect with the Xbox compared to gaming systems that came before it, but also about how the company imagined the player experience through its technology. They offered a premonition about where sensing machines would go in the near future.

To resurrect "you are the controller," Microsoft drew on two cutting-edge areas in its computer science research divisions: computer vision and machine learning. *Computer* or *machine vision* studies how to get computing machines to see. It aims to give computers a visual understanding of a world that to us consists of people, animals, trees, and buildings but to a machine is only numbers. Machines capture the world with their cameras and turn it into the visual language of computing—pixels. These machines then learn to recognize objects in the world by decomposing their constituent parts into primitive lines, edges, corners, and blobs, analyzing the pixel groupings and making educated but probabilistic guesses or *inferences* of what objects the groupings might be.

On the other hand, in *machine learning*, computers improve their abilities to recognize things based on their "experience." But a computing machine's conception of experience is radically different from that of a human. Experience for machines is learning from example, like being able to view images of a thousand cats and then knowing the difference between a dog and a cat if presented with the image of a dog. Machine learning isn't driven by rules or conditional statements like "if x, then y" that tell a system how to react to a possible situation. Instead, it works through a machine being fed thousands or millions of existing examples, called a *training set*, and then, when new data comes in that the machine has not seen before, attempting to make an educated guess about how close or far (the degree of error) that data is from the initial training examples.

One niche area of computer vision research that motion tracking–based games depend on is getting a machine to recognize a human body. Body recognition is a difficult problem for machines to solve and has generated tens of thousands of academic papers devoted to it. But what would happen if the machine could get more realistic information about the difference between a moving body and another moving object in the frame?

The Kinect solved this problem by combining camera sensing with computer vision and machine learning.[26] First, the Kinect's optical sensors came in two varieties and performed two different tasks. A webcam-like camera that detected standard red, green, and blue (RGB) colors was able to "see" a body as it moved in front of the camera and rendered this image on the screen or in the game. But a second set of cameras would generate a wholly different 3D image of a body—one seen not by the gamer but instead by the algorithms themselves.

This second set of sensors generated an image that no ordinary camera could produce—a *depth map* that distinguishes objects in three dimensions. Using a small projector, the Kinect would spray a pattern of invisible infrared dots on the scene that only the second sensor could see.[27] This approach is called *structured light* in machine vision research. The invisible dots that coat objects in the sensor's sight are distorted in size and position when the object moves. The sensor then compares the original pattern of dots to the new distortion based on where the object has moved and calculates the changing depth based on how the dot pattern changes. The process doesn't happen just once—but thirty times a second.[28]

This complex tangle of optical machinery in the first version of the Kinect released in 2010, provided by an Israeli start-up called Prime Sense, was designed for one main purpose: to distinguish a moving body from the background of the scene. But something else also happened. The Kinect used the depth map to build a computer model of the gamer's body in its memory. This was not a trivial task; how to convert moving pixels to something approaching the human body was a long-researched problem in computer vision.

Having to hand-code all of the possible poses that a body makes would have been an almost impossible task, a calculation probably equaling the number of atoms in the universe. Microsoft thus drew on its research divisions' academic knowledge of object recognition and machine learning to match the poses a gamer's body could make with its existing database of moving bodies captured by its film crews around the world. Its algorithms divided the human body into thirty-two different categories—hand, forearm, elbow, top-right corner of the head—and then used probabilities based on the existing training data to make guesses about which body part was which.

This process indeed was a curious phenomenon. Microsoft's sensors and algorithms came to identify your movements and gestures only because they had been trained to recognize the gestures and movements of others. The system could only insert you in the game by recognizing and classifying you based on models representing someone else. In other words, your body and Microsoft's sensors and software had different ways of perceiving the world. Your experience of moving in the game in the moment is one of your whole engaged body, but to the machine, you are only parts, groups of numbers and probabilities.

The Kinect exhibited another curious phenomenon. Shortly after its introduction in November 2010, reports surfaced that the device's sophisticated IR camera discriminated against darker skin. Because the device's IR projector would shine invisible light onto objects (like bodies and faces) and use its CMOS sensor to detect the amount of reflected light bouncing off the object, it seems that Microsoft's engineers hadn't taken into consideration the fact that darker surfaces and skin reflect less light than lighter surfaces. Extensive testing from the US independent nonprofit organization Consumer Reports, however, revealed that the sensor was affected instead by changing lighting conditions since the sensor needed enough light and contrast to determine players' facial features in order to log them into the system in order to play games.[29] We might wonder, however, why Microsoft, whose deep pockets enabled the recording of people's bodies around the world to train its machine learning algorithms for a new era of game play, didn't think to test the Kinect with different skin colors under *changing* lighting conditions. Was this a flaw in engineering or simply a lack of awareness that, in fact, different bodies would be engaged in game play?

The difference between the Wii and the Kinect was between two contrasting paradigms in human-computer interaction: motion sensing versus gesture recognition. Accelerometers, the hardware-based motion sensors of the type deployed by the Wii, highlight the *temporal* aspects of motion and gestures, the fact that they take place over time. After all, acceleration is a measure of the change of something's speed or velocity with respect to time.

In contrast, gesture recognition suggests that there is already something, a gesture or motion, to be *recognized*, allowing for a library of predetermined gestures already existing in the machine's software that can be matched to the newly sensed gestures. While a sophisticated motion sensing or tracking system that can locate multiple points of a body's limbs in space might feed software that can recall an existing model, a gesture-recognition system already understands what a gesture or body is.

This difference is not only technical. It also suggests different conceptual worldviews in how sensing machines relate to us. In contrast to the Kinect, the Wiimote didn't know what a player's body was.[30] It had neither a library of existing bodies to match gestures to nor a model of what skin color the player's body was. Instead, the Wii's accelerometer registered physics and forces: friction, velocity, gravity when the physical world came in contact with it. To work, the controller and its sensor needed to become part of you when you jumped, sped up, or fell fast to the ground. Gravity shifts, speed changes, shock appears. Holding the plastic-encased sensor in your hand suggested that the device was an extension of you. The world didn't exist until the sensor came in physical contact with it.

The Kinect had a different model. All that it needed was for the device's optical sensors and "brain" to detect something in the world in front of the camera and match it with something stored in the machine. In other words, the Kinect saw pixels and ran processes: decomposing, filtering, remembering, classifying, matching, predicting. The physical world went through filters and layers of mediation built into the device until what the Kinect "saw" statistically matched the gamers' movements stored in its memory. For the Wii, the world's physics needed to activate the accelerometer in order for the machine to sense something. For the Kinect, the world was already there. It just had to recognize the right combination of pixels.

Now things have changed. Both controllers are now discontinued. They are technical history, along with the dozens of experimental and innovative controllers with the latest sensors before them that never made it to market in the first place or failed when they got there: fishing poles with accelerometers, plastic bowling balls, ninja swords, a headband to control Atari games using forehead muscles, and a perhaps even more prescient sensor for pandemic times: a pulse oximeter.[31] Announced as the Wii "Vitality Sensor," the device was supposed to have registered the player's pulse and the oxygen amount in their blood through their finger for gaming, but it was never released. Perhaps this was a good thing, given the reports that such sensors indeed are problematic in their "seeing" of different skin colors.

The fact that these sensor controllers were not released, however, doesn't matter. Motion sensing is now culturally engrained. We expect keyboards, trackpads, mice, remote controls, and handheld wireless devices to sense our motion in as transparent a way as possible, without us being conscious of them. Both the Wii and Kinect were successes in working to eliminate the distinction between flesh and pixels in gaming.

Why are these histories of successful and failed sensing machines in games relevant? Because they give a hint of the accelerating shape of things to come. Computer game play is now manifested through sensing machines. It is also marked by accumulation. More machine learning, more sensing, more audio/visual fidelity, and even more data equals more experience. Sensing machines thus make play both exceedingly real and increasingly invisible.

But sensing in games is still trying to recover the dream of transparency. The technologies should fall into the background as experience comes forward, deeply mediated but uninterrupted. Seamlessness and immersion become the call of the day, riding a precarious line between awareness of the machinery and forgetting it. To paraphrase artist and *Happenings* inventor Allan Kaprow, as sensing machines, games thus move from life-like art to art-like life.[32]

II Immersing

4 Borderless

In the end, you are tired of this old world.
—Apollinaire, *Alcools*

If you take the Yurikamome elevated subway near Tokyo Bay, you soon enter Odaiba, a chunk of reclaimed land filled with futuristic hotels, amusement centers, and office buildings. Amid a Toyota showroom in a shopping area called Palette Town, there is a metal-clad building—the MORI Building Digital Art Museum. From its drab outside, one would never expect this "museum without a map" would hold something verging on the spectacular: the Tokyo-based art collective teamLab's permanent, multiroom media spectacle called *EPSON teamLab Borderless*.[1]

This 10,000 m² artwork leaves me almost speechless when I visit in early March 2019. After paying a thirty-five-dollar entrance fee, I enter into an enormous labyrinth of rooms. Every surface, wall, floor, and ceiling is bathed in computer-generated projections of animated butterflies, waterfalls, blooming flowers, and swarming birds. A space with mirrored floors called Crystal World is filled with clear strands of some three hundred thousand LEDs that continually pulse, burst, and shimmer, producing mesmerizing waves of light.

As I walk from one Instagrammable environment to another, the impressions are fleeting due to the sheer number of stimuli. From every possible direction, images overtake the walls and floors. Wrap-around mirrors push the rooms into a feeling of infinity, giving us as visitors the sensation of feeling small. Hundreds of spiraling light beams produce a sense of vertigo. English translations from the Japanese titles that accompany the different

Figure 4.1
Light Vortex II, teamLab *Borderless*, Tokyo. Copyright teamLab.

installations awkwardly attempt to convey *Borderless*'s totalizing media experience in words. In *Universe of Water Particles on a Rock Where People Gather*, a massive animated waterfall and sloped floor resembling a rock swallows up visitors. *Flutter of Butterflies beyond Borders, Ephemeral Life Born from People* contains flocks of animated butterflies that burst forth on the walls and fly between the rooms.

New age electronica synthesizer music accompanies the ever-transforming digital forms laid onto the real architecture. As I wander dazed in the din, the music seamlessly merges from one track into the next, bombarding my ears with a mélange of sounds, from temple bells to schmaltzy film soundtrack strings. It's hard to fathom how anyone could produce a sound-space so dense given that *Borderless* is based in Tokyo itself—a place that probably has more injected artificial sound than any other megalopolis in the world—but the experience crammed with audiovisual effects and visited by 2.3 million kids and adults in 2019, its first year of operation, succeeds.[2]

How is this wondrous media hallucination in the midst of Tokyo created? The magic ingredients are impressive: 520 networked computers, 470 EPSON video projectors, and hundreds of sensors.[3] Dozens of cameras capture the spectators' movement and use it to partially trigger audiovisual events. Similar to standard home security systems, the cameras, together with infrared LEDs that only the camera sees, serve as motion sensors, detecting and tracking how much heat visitors give off. The cameras are like those that have long greeted us in Asian airports (and, after the 2020 COVID-19 pandemic, in airports around the world), which measure the temperature of newly arriving visitors to check whether they are feverish.

The triggered animations and effects pull the visitors deeper into the spectacle, transforming them from passive onlookers into active participants. If I near the walls or step on particular areas of the floor, the cameras detect this and the payoff comes: the ability to influence the quality, density, or action of the animate projections that fly over the surfaces of the spaces.

To enable *Borderless*'s visitors to influence the media landscape, the cameras and their software use sophisticated techniques derived from computer vision. As we saw with the Kinect, computers can't comprehend anything but numbers—in the case of computer vision, how one pixel is lighter or darker than a neighboring one. By searching through these values, the software can detect more complex shapes or patterns. The number of techniques for getting machines to see is a little overwhelming at first. *Motion tracking*, for example, uses a camera and software to calculate the difference of pixel colors between frames of video, to get a sense of how motion changes over time. *Blob detection* seeks out groupings of similar pixels called *blobs* in an image that has a different size or color than what surrounds them. Another method, called *edge detection*, tries to determine the edges or corners of an object from the order of pixels in order to distinguish an object's boundaries.[4]

But there is one fundamental thing the sensors don't know: the significance of a human body and what it can do. The images *Borderless*'s cameras capture are not meant for human perception but for software algorithms. What is sensed ignores most of the features we associate with being human. The algorithms can't measure the degree of joy or confusion we experience as we lose ourselves inside *Borderless*'s maze of sensory impressions. They don't know what our mood is when we enter and later when we exit. They don't know what histories we carry into the room, what physical ailments we might have, what emotions we may harbor inside. For all intents and purposes, to these camera sensors we are simply groupings of brighter or darker pixels, calculated based on ever-changing frames in the ceaseless flow of images that rush into these devices.

⌒

Borderless is the artwork as sensing machine tout court. It folds us visitors into its operations, expanding our senses and leaving us awed and numbed in the same breath. It seeks to give the impression that we have agency, an active hand in creating a unique individual experience for others as well as ourselves. We move and the computer-generated butterflies swim around us. We step into a zone and are greeted by unfolding Chinese characters and digital watercolors. We near the wall and a parade of animated chariots follows our shadow. In other words, the capture of our presence is necessary for the machines to work, and our participation tightly couples us to them. The customer-visitor is the new creator in an event specifically designed and engineered to leave us borderless.

But despite its ambitions to take us to the extremities of our experience, *Borderless* still has limits—those of the human senses. The environment's sensing and computing technologies are undeniably cutting-edge, orchestrating a spectacle that reaches an almost unprecedented level of seamlessness and precision, comparable to the control bunkers that run smart cities or Disney World. The number of electronic sensors and computational firepower used to create teamLab's spectacle is larger than one can count, but ironically, there may be less room for our fragile human sensors to process it all.

That an artistic experience like *Borderless* is only made possible by sensors and computers but entirely dependent on our human sense experience for its success is not surprising. "Technologies," wrote media scholar Marshall McLuhan in his 1964 essay "The Medium Is the Message," are "extensions of the senses." For McLuhan, the senses could be seen essentially as technologies and vice versa. "The effects of technology do not occur at the level of opinions or concepts," he stated, "but alter sense ratios or patterns of perception steadily and without any resistance. The serious artist is the only person able to encounter technology with impunity, just because he is an expert aware of the changes in sense perception."[5]

While the arts, particularly in the twentieth century, had always dealt with the senses, McLuhan argued that the current technological conditions (at the time, the 1960s) were suggesting ever-new registers of sense perception that would be coupled with the coming technological shifts of computers and networks to extend the nervous system. This vision of a new kind of participation made manifest by being inside an experience rather than exterior to it rapidly took off.

During the same period, visual artist Allan Kaprow sought to tear down the distinction between the artistic object and the viewer with his time-based artistic events called *Happenings*.[6] Kaprow distinguished the immersive arts from the arts of painting and sculpture when he announced, "go IN instead of LOOK AT."[7] Sixty years later, immersion is not only a technique but also a brand: a novel modus operandi of our ever-new spectacle culture that spans art museums, shopping malls, theater performances, and live gaming environments. Artworks no longer consist of objects on walls but instead, as musician Brian Eno states, "triggers for experiences."[8]

Borderless perfectly demonstrates that the interplay between immersion and participation enabled through sensing machines creates a new cultural normal. Along with teamLab, similar immersive formulas are being tested by international creative organizations with names like Meow Wolf, Dreamscape Immersive, Moment Factory, Marshmallow Laser Feast, The Phenomena, Superblue, and Punchdrunk, among myriad others.[9] Whether conscious of it or not, these new organizations have adopted a well-known playbook for their strategies.

In a now seminal article that appeared in the July 1998 issue of *Harvard Business Review*, entitled "Welcome to the Experience Economy," business consultants Joseph Pine and James Gilmore describe how experience, "the staging of memorable events by a company," should be pursued as a novel and "distinct form of economic output."[10] Pine and Gilmore trace the *ur-experience* economy back to Walt Disney, who carved out a new brand empire in the mid-twentieth century by vertically integrating Mickey Mouse, Donald Duck, and other cartoon characters across films, television, toys, theme parks, and, years after his death, video games, Broadway musicals, and even cruise lines with their own Caribbean island destination.

The new creatives involved in mastering the rules of the experience economy have pioneered the conception and production of location-based entertainment extravaganzas that blur art and entertainment and quickly replace older forms of spectacle culture, like cinema or television, with computer-assisted interactivity. The goal of immersion is now to transfix visitors by producing a form of experience in which it is almost impossible to tease out the distinction between reality and technologically orchestrated fantasy. "When it feels this real, is it virtual or reality?" asks Dreamscape Immersive, a start-up comprised of Disney Imagineering experts and Swiss VR technologists. It uses the phrase "team up, gear on, leap in" to describe its social VR experiences.[11]

Combining theme park effects like motion platforms, artificial wind, and water sprays with cinematic scale VR, Dreamscape created its first location-based entertainment experience called *Alien Zoo* in 2016 in an upscale Los Angeles shopping mall. *Alien Zoo* involves six audience members at a time, who undergo a forty-minute VR experience that transports the group to an alien planet of animated creatures for a cross between *Jurassic Park* and *Finding Nemo*. Not satisfied with just LA, *Alien Zoo* is now franchised to other cities, including the epitome of retail experience, the Mall of Dubai.

Meow Wolf is close behind. The Santa Fe–based art collective has become an entertainment conglomerate with the success of a 20,000 sq. ft., immersive, interactive environment called *House of No Return*. With a Burning Man meets Disney on magic mushroom aesthetic, this once-poor group of artists is currently the anchor tenant for AREA15, a monstrous Las Vegas–based "immersive experiential retail design" complex that seeks to reinvent that now failed twentieth-century vision of immersive participation: the shopping mall. Planning major permanent installations in Denver and in Washington, DC, as well as an interactive hotel in Phoenix, Arizona, Meow Wolf's combination of art, immersion, and commerce led one curator to critically write that the group's work was "a supreme act of late-stage capitalism disguised through the collective's mantra of the underdog as art savior."[12]

How then do we understand these phenomena? Blurring the lines between immersion, art, retail, and experience is not an easy thing. It wholly depends on instilling the belief that visitors ("guests," in Disney speak) are not separate from the environment but inseparable to its operation. Using location, touch, and motion sensor data to both feed computer-based interaction and to gather valuable user data for targeted advertising, it is also not difficult to see why sensing systems are so necessary for these new immersive forms. They help weave human visitors' presence into sets of computer instructions that are designed to switch on and off sensory experiences, ranging from the playful and whimsical to the frightening in a gigantic loop between the space and the visitors' bodies. Sensing dials up the intensity of the media, using the public as the trigger.

While created with the latest digital projectors and computers, *Borderless* actually has some historical roots. It harkens back to an older form of spectacle called the *phantasmagoria*, an exhibition concept emerging at the tail end of the eighteenth century in Europe and America. The phantasmagoria featured optical illusions of ghosts and phantasms projected onto the walls of darkened rooms by means of a new technology of the time—the magic lantern. The magic lantern was an early cinema precursor consisting of a container with a lamp and lens-like aperture used to project and magnify images from painted slides, glasses, or (later) film onto surfaces. The phantasmagoria was by no means a purely visual spectacle. It would also be accompanied by ghostly soundtracks consisting of bells, wind, whispers, and thunder.

But the phantasmagoria is not the only reference for teamLab. In fact, the group's installations running in Tokyo, Singapore, a Macau casino, and museums around the world liberally borrow from the entire history of immersive spectacles: artificial environments produced with technological means the aim of which has been to surround, envelop, and saturate our senses through media. A room of more than one hundred glass screens is inspired by the nineteenth-century Pepper's ghost technique in order to create the impression of floating images of moving creatures; floor-to-ceiling projections recall early surround IMAX films and the spectacles of bygone world's fairs; and chambers with wraparound mirrors evoke the "infinity rooms" of the eminence grise Yayoi Kusama—immersive artworks by the Japanese artist that fill contemporary art museums around the world where visitors wait hours to spend one minute inside a tiny reflective space that evokes the cosmos.[13]

In other words, while indisputably impressive in terms of its scale and technical implementation, *Borderless* is really not all that new. Already at the dawn of the twentieth century, artists and creators of all stripes in Europe and later America, Asia, and South America were laying the groundwork for what would become teamLab's reality one hundred years later. They imagined spectacular environments that would erase the line between passive watching and enveloping participation, and the new machines of their time—the airplane, the railroad, the radio, the hydraulic press, and the cinema screen—were their guides.

The vast majority of these visions were fantasies. The technologies of electronic sensors and artificial intelligence that make up the sensing machines of today were nonexistent at the time, and artists would have to wait another fifty years for the technical means to make their dreams happen. But it wasn't that sensing machines in the hands of artists were unimaginable. In the 1938 *Exposition internationale du Surréalisme* at the Gallerie des Beaux-art in Paris, the controversial artist Marcel Duchamp, who worked on the event as an early curator-exhibition designer, imagined using "magic" or "electric" eyes to heighten perceptual vertigo in his audience. Duchamp hoped to reveal the paintings in the exhibition (which would have been left in the dark) by solely triggering the lights with the sensors in order that traditional viewing at a distance would be destroyed, a concept that went ultimately unrealized.[14] By plunging the audience's vision into pitch darkness, Duchamp's attempt, as well as future ones, was the solution to what the artist famously criticized as the dominance of "retinal art."

The artists and designers of Weimar Germany also envisaged sensing machines, particularly Bauhaus artists such as Hungarian-born László Moholy-Nagy. In addition to his famous photograms and graphic design, Moholy-Nagy worked as a stage designer in the 1920s, conceiving of new forms of multisensory theater. In an unrealized concept entitled *theater of the mechanical eccentric* from 1924, Moholy-Nagy proposed a new kind of "theater of totality," utilizing the latest technological machinery to create what the Hungarian émigré called "a synthesis of form, movement, sound, light (color) and scent." Moholy-Nagy sought a new fusion of machine and organism through a subordination of human actors to the power of technology. He wanted the audience to "fuse with the actions of the stage at a moment of cathartic ecstasy" by using dynamic light, sound, and "electric bodies" to create what he called "great clashing masses or accumulations of media."[15]

How were artists so foresighted about what was to come? Even as the economic conditions were dire and the political climate unhinged in a shell-shocked, post–World War I Europe, creators sought both aesthetic and spiritual salvation through the tools of their time. Most extraordinarily, they envisioned projections in the sky or moving, machine-like theater spaces when, as American art critic Robert Hughes wrote of post-1917 Russia, "there was hardly enough surplus wattage in all of Moscow to run an egg timer."[16]

Yet in a Europe grappling with the legacies of colonialism and the collapse of empire, the longing for renewed visions of the human under new political models like socialism fused with rapidly developing industrialization was a powerful force to both adhere and react to. In the hands of creators, technology was no mere instrument; it was a transformative tool. "The era that humanity has entered is an era of industrial development," wrote Lyubov Popova, one of the most important women artists in post-revolution Russia, "and therefore the organisation of artistic elements must be applied to the design of the material elements of everyday life, i.e. to industry or to so-called production."[17]

But the immersive arts also had another goal. They were seen as a potent means for enabling a spiritual quest into mystical, altered states of consciousness that, so it was argued, could only be achieved through the new forms of interaction and immersion, the creation of extreme aesthetic encounters in which one felt part of the world rather than a stranded, distant onlooker. There is no clearer expression of this sentiment than when French actor and later dramatic theorist Antonin Artaud announced in 1932 a "Theater of Cruelty," a new kind of theater that "would act directly and profoundly upon the sensibility through the organs" and where "we abolish the stage and auditorium . . . where the spectator, placed in the middle of the action, is engulfed and physically affected by it."[18]

Thus, in the run up to revolution and the coming great wars, grand and ultimately unrealized utopian forms of spiritual awakening began to emerge, like Russian composer and supposed synesthete Aleksandr Nikolajevič Skrjabin's *Mysterium*, an unrealized, seven-day-long event with the lofty goal to fuse music, dance, theater, and architecture with all of the senses into a complete experience that would take place in a specially built space at the foot of the Himalayas.

Visions of machines and environments extending and blurring the lines between spectating and participating through the senses had another side as well. Writing at the verge of the second destruction of Europe, the twentieth-century German cultural critic Walter Benjamin reacted to the fervid transformation of sensory perception due to new technologies. "The manner in which human sense perception is organized, the medium in which it is accomplished," Benjamin wrote in a famous 1935 essay on the transformation of the artwork's role in an age of technical reproduction, "is determined not only by nature but by historical circumstances as well."[19] The historical circumstances that Benjamin was alluding to were the emergence of the cinema, a medium in which there was no longer a reality where the senses could be separated from technical life.

The film camera was the sensor of Benjamin's time. Described by him in an almost animistic manner, the camera captured and revealed a hidden world that human vision ignored or wasn't capable of seeing. Its "unconscious optics" would arrest the data of modern life, expose the "hidden details of familiar objects," expand notions of space and movement, and, through slow motion, shatter the experience of time. The camera created a new kind of political art form for the masses, together with an equally transformed observer, one whose reality was dictated by the new technology.[20]

Ultimately, this unparalleled mediated technological experience at the birth of motion pictures and the industrial transformation of urban life was one of shock. In an era when "technology has subjected the human sensorium to a complex kind of training," Benjamin foresaw that the only response to such an amount of overwhelming sensory input would be to shut down. "The greater the share of the shock factor in particular impressions, the more constantly consciousness has to be alert as a screen against stimuli; the more efficiently it does so, the less do these impressions enter experience," he wrote four years after his analysis of the artwork in the age of technological reproducibility.[21]

On March 28, 1935, Walter Benjamin's thesis on training the senses for shock would be put to the test. On that day, German filmmaker Leni Riefenstahl's notorious Nazi propaganda film *Triumph of the Will* made its much-heralded cinematic debut. As the Nazi's commissioned chronicler,

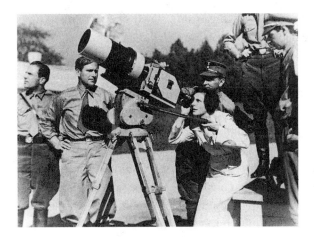

Figure 4.2
Leni Riefenstahl filming *Triumph of the Will*, 1934.

the thirty-two-year-old Riefenstahl would advance the film camera's sensing possibilities to dizzying new aesthetic and political heights.

Triumph of the Will's 114 minutes portray a singular event: the buildup to and occurrence of the Nazi party's monumental 1934 Nuremberg Convention. Riefenstahl's technical apparatus of thirty cameras glides with a God's-eye view over the choreographed spectacle of the gathering. Watching the film in 2020, the cameras almost seem to predict our drone-filled present. They are relentless, incessantly in motion, cutting between extreme close-ups and distorting wide angles while sweeping over seven hundred thousand gathered soldiers, party officials, and interested onlookers. The cameras effortlessly swap human bodies with the architectural features of the massive parade grounds where the rally took place: faces morph into swastika-adorned banners, architectural elements become the sky, human bodies convolve into choreographed rows of automata.

Later, from his distant exile in the United States, German cultural critic Siegfried Kracauer perceptively wrote that the film's "sumptuous orchestration employed to influence the masses" had achieved a complete "metamorphosis of reality" through technology.[22] As Nazi propaganda czar Joseph Goebbels claimed that propaganda was a creative art, *Triumph of the Will's* cameras sensed and produced an unprecedented combination: an aesthetic of underlying unease and slowly encroaching terror coupled with an ecstatic depiction of extreme order.

Walter Benjamin's analysis of the sensing power of the camera to capture mass movements over the limitation of the human senses, the naked eye, was not only aimed at Hollywood. He argued that "mass movements, including war, constitute a form of human behavior which particularly favors mechanical equipment."[23]

Technology's role in altering the senses and perception itself was a complex and contradictory matter. For Benjamin, it could become a liberating tool for the masses but could also easily be put in the service of authoritarianism by introducing "aesthetics into political life."[24] Riefenstahl's sensing camera forever burned into celluloid a record of the coming disaster that came, as Benjamin wrote, because "technology has not been sufficiently developed to cope with the elemental forces of society." Thus, with spectators transformed into participants and under shock from the increasing media manipulation of daily experience, a new paradigm emerged in which sensory immersion would become inseparable from political power.

The modernist pre–World War II European avant-garde acutely demonstrated that in the context of the arts of sensorial immersion, sensing machines were as deeply intertwined with political forms of power as they were with aesthetic transcendence. The political forces associated with immersion and participation in the arts certainly did not abate after the destruction of Europe with the technologies of war. They were transformed as the imagined technologies of the prewar world began to materialize in the Cold War–dominated, technoscience one, in which sensors would play an increasing role.[25]

The postwar cultural imaginaries of immersion and participation took place against the backdrop of two seemingly unrelated but connected transformations: a relentless drive for computer-based systems focused innovation as a key factor in winning the Cold War, and a counterculture emerged in which artists sought to fulfill a spiritual quest for self-transformation and self-alteration by way of these very same machines. New technical jargon that had been the territory of mathematicians and engineers—information, feedback, communication, control, behavior, and interaction—entered artists' lexicons, and they quickly sought techniques to experiment with these concepts among multiple publics.

Growing international movements in conceptual art brought artists to seek out a new "emancipated spectator" by way of machine-human interaction that their earlier twentieth-century forebears only could have dreamed of.[26] Already in the 1950s, writings emerged announcing "Cybernated Art" (Nam June Paik), *Arte Programmata* (Programmable Art), and "Systems Esthetics" (Jack Burnham).[27] Artworks were envisioned no longer as static objects that would sit motionless on walls and pedestals. They would be reborn into new kinds of polysensorial spaces that relied on crude sensors capturing the movement and presence of their observers in order to cause changes in those spaces. The arts and, in effect, the larger counterculture soon became a vast laboratory to experiment with new techniques aimed at creating coreciprocal, feedback-based relationships among the artistic object, its observers, and the environment.

Establishing these coreciprocal relationships through newly appearing electronic technologies was not a trivial matter. In the mid-1950s, electronic technology was still crude and prohibitively expensive, with the integrated circuit, which would compact electronics onto silicon chips, still to be invented (in 1958) and portable computers decades away.

Sensors and rudimentary electronics used in artworks were profoundly unstable—essentially bundles of exposed wires and components that were hooked up and troubleshot minutes before exhibition openings in prestigious international art festivals like the Venice Biennale and Documenta. Indeed, given these limitations, it seems almost crazy that groups of artists became interested in using sensors as the basis for their work.

But the desire for aesthetic self-alteration through sensing machines ran deep. Visual and performing artists and engineers reached for whatever they could get their hands on to produce new "interactive" art works—what one Italian collective, called Gruppo T, dubbed *works in becoming*. The goal in these emerging art experiences was to create open-ended, "unfinished" works by employing simple detection devices—usually sensors used to discover the presence of an object by using a light transmitter, called *photoelectric cells*—to effect some change in the artwork.

Essentially like a light switch, the visitor's motion would block the path of light between a sensor and a receiver, thereby changing the sensor's electrical voltage. This change would then be "interpreted" directly in hardware and result in triggering media: turning lights on or off, spinning motors that would rotate mirrors or prisms or colored gels to shift the color in a space, or starting sound, all of which aimed to give a sense that the artworks were "responding" to visitor presence.

Sensors not only allowed artists to glean something about what was going in their newfangled "installations," chockablock with psychedelic light patterns, strange sounds, and moving parts. These technologies would also enable audiences to potentially understand how they were part of a much larger system of feedback processes within the artificial and, sometimes, natural environments they inhabited.

American artist Robert Rauschenberg heralded this new age of "reactive environments," and in the United States he was one of the leaders in helping reimagine what artistic experience under the influence of sensing machines might be. Already established as a leading visual artist by the mid-1960s with his "combines," collage-like mixed-media works that confused the line between painting and sculpture, Rauschenberg had long been interested in incorporating the visitor's presence and perception into his work. From 1962 to 1969, and in principal collaboration with Billy Klüver, a Swedish-born electrical engineer who was working at Bell Telephone Laboratories's suburban-sequestered New Jersey research center, Rauschenberg developed four room-scale mixed-media environments that "would be responsive to weather, to people viewing it, to traffic, noise and light."[28]

Rauschenberg's installation *Soundings* (1968) directly explored the potential of sensor-activated technology to both reveal and confuse the confines of self-image. Standing in darkness before a thirty-six-foot-long wall of multiple layers of Plexiglass panels with silkscreened images of chairs printed on their back surfaces, viewers first glimpsed themselves in a mirrored surface that constituted the first layer. Through hanging, voice-activated microphones that functioned as sensors, Rauschenberg encouraged the visitors to react to the work by yelling, singing, and talking—what he termed a *one-to-one* response.

Filtering the sounds of children screaming or adults in conversation for different frequency components, the visitors' voices activated a series of hidden, high-intensity spotlights that revealed and hid the back-printed layers while blurring the viewers' reflections. The sensing brought the faces

Figure 4.3
Robert Rauschenberg, *Soundings* (1968), mirrored Plexiglas and silkscreen ink on Plexiglas with concealed electric lights and electronic components, 96 × 432 × 54 in. (243.8 × 1097.3 × 137.2 cm), Museum Ludwig, Cologne, Ludwig Donation © Robert Rauschenberg Foundation / VAGA at Artists Rights Society (ARS), New York / SOCAN, Montreal (2021).

and bodies of the visitors' reflections in and out of focus, blending them into the printed surfaces. The object and the self would become one.

But it was in *Mud Muse*, a 1968–1971 project made under the auspices of the experimental Art and Technology program organized by the Los Angeles County Museum of Art (LACMA) and its young curators Maurice Tuchman and Jane Livingston, that Rauschenberg could fulfill his desire to create the ultimate interactive environment, "which would also be responsive to forces outside the jurisdiction of the viewer."[29]

In a new move by the artist, *Mud Muse* dispensed with the viewer's presence entirely while continuing to amplify the notion of the artwork as a systems-oriented sensing and feedback machine. Rauschenberg, along with engineering collaborators from the LA-based Teledyne Technologies corporation, created a gigantic, closed-loop environment: a nine- by twelve-foot glass tank filled with a thousand gallons of drillers' mud and activated by a sophisticated, custom-engineered series of sound-activated pneumatic pumps that would cause the mud to boil and burst like a geyser and respond to its own behavior.

Mud Muse was a curious work for Rauschenberg. It was his second art and technology project in a little over three years that involved an odd set of collaborators for artists at the time: engineers from corporate research environments. Teledyne, Rauschenberg's engineering collaborator on *Mud Muse*, to which he had been connected through the LACMA initiative, would not necessarily have been the first choice for creators with a sociopolitical conscience in the 1960s countercultural climate of the Vietnam War. A behemoth technology conglomerate with a name that meant *power from afar*, Teledyne served as a major defense contractor developing some of the first unmanned drones for bombing missions during Vietnam and in the mid-1970s pioneered the buyback of its own stock to inflate its falling share price.[30]

In retrospect, the LACMA initiative as a whole seems more than problematic. It was initially designed by Tuchman and Livingston as a new model to support the burgeoning art and technology scene by directly brokering relationships between artists and corporations, but in the end its first artist roster was a narrow pick of twenty-six well-known, all male and, with the sole exception of the Black sculptor and engineer Frederick Eversley, all white artists such as Andy Warhol, James Lee Byars, and Claes Oldenburg.

The Art and Technology program also danced with the corporate devil in the name of aesthetic innovation. Teledyne was in good company with forty other participating corporations in the project, mainly from the LA-based military-industrial complex.[31] Certainly, the visual art world didn't think kindly of the initiative in this respect. The inclusion of RAND (originally, a research project of the Douglas Aircraft Corporation whose sole client was the US Air Force and that eventually became a private think tank), the Jet Propulsion Laboratory, Hewlett-Packard, Rockwell, Lockheed, Garrett Corporation, Litton Industries, and others lead Max Kozloff, a leading critic, to argue in the journal *Artforum* that the involved corporations had been "subsidized decisively by the American government . . . through the business of slaying." Kozloff's vindictive continued: "While the convulsions of the 1960s [Vietnam, race riots, the assassinations of Robert Kennedy and Martin Luther King, Jr] were taking place, inflaming the radicalism of our youth and polarizing the country, the American artists did not hesitate to freeload at the trough of that techno-fascism that had inspired them."[32]

It was through his association with engineer Billy Klüver that Rauschenberg instantiated his relationship with another major corporate collaborator: Bell Laboratories. While Bell Labs was busy inventing the tools of the digital twentieth century—the bit, speech recognition, the transistor, cellular phone technology, and a mathematical formula called the fast Fourier transform (FFT) that would eventually be incorporated into every digital device in our households—Klüver and a group of somewhat "rogue" engineers were also developing a radical art program in collaboration with Rauschenberg and artist Robert Whitman.

Nine Evenings of Theater and Engineering was a famous experiment linking emerging and established New York downtown performing artists (with a higher percentage of women artists, unlike the LACMA initiative), such as Yvonne Rainer, Lucinda Childs, and Deborah Hay, as well as John Cage, David Tudor, Alex Hay, and Robert Whitman, with some thirty engineers from Bell Labs' New Jersey headquarters to create innovative theater, dance, and visual art–based performances using state-of-the-art research technologies. Originally organized for a Swedish art and technology festival that fell through, the artists and Klüver's team eventually spent 1.5 years and an estimated 8,500 hours of work on the creation of a series of performance works.[33]

More than ten thousand spectators jammed into the Park Avenue Armory building in New York City in October 1966 to witness the future of the performing arts. *Nine Evenings* was a panoply of just-invented sensing technologies and electronic gizmos deployed in the name of avant-garde experimentation. The prevailing objective was twofold: to create new aesthetic terrains by way of artist-engineer collaborations, and to humanize dehumanized means by putting the rigid, quantitative, and control-based framework of engineering technologies into playful contexts.

A project by minimalist choreographer Lucinda Childs employed Doppler-based sonar to detect the random movement of dancers and objects and used this information to control the actions of light and sound without being directly scripted by the artists. Rauschenberg's massive *Open Score* performance used tennis rackets reengineered with wireless FM transmitters and IR-sensitive television cameras to sense five hundred performers in the dark. John Cage's chaotic musical event *Variations VII* involved banks of photoelectric cells triggering groups of lights while contact microphones, Geiger counters, an early set of worn electrodes to detect brainwaves, and open telephone lines were explored as devices aimed to capture all sounds "in the air at the moment of [the] performance."[34]

Rauschenberg's personal forays into art and technology and his founding, together with Klüver, Whitman, and engineer Fred Waldhauer, of the long-running nonprofit organization Experiments in Art and Technology (E.A.T.) even before *Nine Evenings* burst onto the scene didn't particularly sit well with a humanist-centric art world. One critic declared that Rauschenberg's *Soundings* "reduced the human agent" to essentially a "switch"—a criticism that would be continually dished out by hostile visual art critics as sensing technologies were rapidly incorporated into artistic works.[35] These critics feared that the traditional experience of art was being undermined by making the viewer ever more necessary as part of the overall aesthetic experience.

If mainstream art critics had a field day labeling *Nine Evenings* "a disaster" or the LACMA project a "million-dollar boondoggle," these events also revealed deeper anxieties. How was it possible that experimental artists seeking countercultural consciousness would employ technologies destined for human annihilation in their work? Within the democratic context of the postwar US, the LACMA- or E.A.T.-sponsored projects were, after all, far beyond the context of Leni Riefenstahl's whitewashing of fascism by aestheticizing politics with the technologies of cinema.

Artists and a handful of sympathetic curators using the new sensing and electronic technologies were not naive to these anxieties, particularly acknowledging in later interviews the ethical ambiguities in using such tools under the pretense of democratizing experience.

But a more striking criticism arrived from light artist James Turrell. Together with visual artist Robert Irwin, Turrell collaborated early on an unrealized LACMA Art and Technology project with Garrett Corporation psychologist Edward Wortz to explore what they labeled the *sense of sensing*.[36]

Turrell, Irwin, and Wortz's research aimed to create a sensory chamber that sought to directly manipulate the conscious state of the visitor by combining two experimental settings used in psychophysics research: the *anechoic chamber* (an acoustically treated room that removed acoustic reflections) and the *Ganzfeld effect*—the simulated equivalent of flying into a bank of fog where no borders or edges exist in the visual field for a perceiver.

The chamber idea was gradually abandoned, but Turrell and Irwin's project did not drop their interest in transforming the self by way of sensing and feedback. The artists and Wortz began a series of "alpha wave" conditioning experiments—measuring brainwaves to bring subjects into meditative and altered states of consciousness. The project soon came to naught. Turrell dropped out, while Irwin and Wortz continued to work on other art and science collaborations for years later. But according to Turrell, the entire painstaking effort to use advanced technologies for self-transformation was primitive, at best: "We have devices, sensors, alpha conditioning machines. The machines are just manifested thought. Technology isn't anything outside us. . . . We just go about it very clumsily and very wastefully. Because we have to actually make all these devices, we have to go to the moon, we can't see the cosmos in a rock, and we can't meditate without having this thing strapped on us."[37]

Turrell's comments seemed to hit the Pause button on the belief that self-transformation could be achieved through Pavlovian techniques that used sensing machines to measure and condition self-transformation.[38] Far from creating new forms of selfhood, these technologies essentially turned visitors into psychological guinea pigs under the guise of art. But the criticisms ended neither the belief that sensing machines would bring about transformations nor the desire for the technique of feedback to accomplish it.

⌒

As part of our current daily lives filled with ubiquitous automation and 24/7 technological saturation, teamLab's sensory transformation through immersion in *Borderless* that began this chapter forms one kind of imaginary. In comparison, however, in the mid-1960s, another very different understanding and vision of immersion and participation appeared. In 1965, the pioneering feminist theater director Joan Littlewood teamed up with an experimental avant-garde architect named Cedric Price and eccentric psychologist and engineer Gordon Pask to create a new kind of sensing machine: a "university of the streets" that they labeled the *Fun Palace*.

Seeking to build what the creative team called "a laboratory of pleasure," the *Fun Palace* was to be a colossal flexible architectural structure:

a many-leveled environment consisting of moving walkways, escalators, and ephemeral architectural materials like vapor and light curtains, which would be capable of continually rearranging itself in response to the events that would take place within. How a flexible yet built environment imagined to accommodate fifty thousand visitors at a time would be able to change its architectural form on a moment's notice remained a mystery, one that Pask was determined to solve.

Pask, who had gone from designing sensing machines like a futuristic musical instrument called the *MusiColour* that produced colors in relationship to electronic tones to conceiving of teaching machines and chemical computers, was the ideal collaborator.[39] He would allow Littlewood and Price the possibility of reconceiving their playful vision to be emblematic of feedback through the newly emerging scientific paradigm of cybernetics.

The political belief that sensing machines could create a new kind of self through adaptive cycles of feedback did not come out of nowhere. Feedback was *the* central concept in the burgeoning interdiscipline of cybernetics, of which Pask was part. The founder of cybernetics, American mathematician Norbert Wiener, defined it as "the entire field of control and communication theory, whether in the machine or the animal."[40] Cybernetics aimed to study the ability of machines, whether biological or mechanical, to steer their own behavior through feedback loops that would occur through sensing and effecting (i.e., changing) their external environments. Information and messages sent between the components of a system would establish structures of organization and enable the system to organize and self-regulate its behavior without influence or control from the outside.

While principally a scientific domain entangled with the military-corporate invention of the period that led to technological breakthroughs in areas such as signal processing, machine automation, and robotics, cybernetics was quickly adopted into nonscientific contexts as well, from running economies and designing the flow of cities to making art.[41]

Because Wiener modeled cybernetics on the neurophysiology of sensing and effecting systems that could adapt their actions though feedback, its framework of feedback and response supplied scientists like Pask with the conceptual and technical tools to create new kinds of sensing machines. But cybernetics also provided a curious political context, one delicately balanced between freedom and control that, in contrast to the aesthetic liberation that its philosophy suggested, would pose larger and unanswered questions around the issues of governance and regulation that the *Fun Palace* sought to experiment with in real life.

Pask closely collaborated with Littlewood and Price to develop the *Fun Palace* as a gigantic machine at architectural scale based on core cybernetic principles like sensing, feedback, control, self-regulation, and learning. He brought to the project prodigious knowledge of feedback modeling, control systems, game theory, and operations research that would be used to create the *Fun Palace*'s "games and tests that psychologists and electronics engineers now devise for the service of industry or war."[42] Even more importantly, Pask saw the project as something beyond fun. It would be a more profound environment for "transcending the familiar."[43]

With Pask's collaboration, the *Fun Palace* was thus rethought from a responsive architectural environment to an algorithmic modeling system. Sensors were imagined to capture visitor flow patterns, group movement, and even "what is likely to induce happiness." Obscure mathematical disciplines (at the time) such as decision theory and predictive modeling would model this data. The feedback cycle was foolproof. If the input was that of *unmodified people*, the output of the system would be more extraordinary: *modified people*.[44]

Was Littlewood, a radical Marxist theater maker of antiwar agitational propaganda plays, aware of the possibilities of new forms of political and social control that might emerge from the *Fun Palace*'s cybernetic foundations? Clearly. Writing in a 1968 manifesto called "A Laboratory of Fun," she not ironically claimed that "the *Fun Palace* would be a foretaste of the pleasures of 1984," referring to George Orwell's dystopian future vision of total surveillance.[45] The pleasure and control enabled by technologies would not just go hand in hand. They would be necessary bedfellows for the new and radical forms of experience that the *Fun Palace* as a sensing machine for aesthetic pleasure and social transformation sought to provide the public.

As you might have imagined, the *Fun Palace* was never realized. Although financing had been secured and a site in central London identified, the inevitable complexity of a project stretching over multiple years eventually made it untenable. Architects Richard Rogers and Renzo Piano claimed that their controversial 1975 Centre Pompidou in Paris was directly inspired by Littlewood, Price, and Pask's vision, but the comparison is weak. The Pompidou is a standard museum, whereas the *Fun Palace* was conceived to

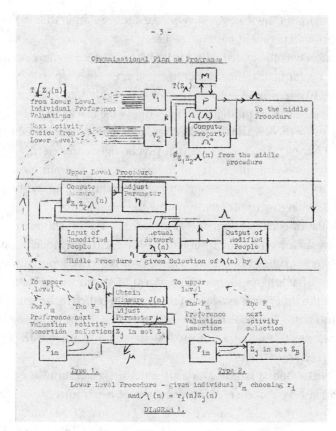

Figure 4.4
Gordon Pask, "Organisational Plan as Programme," from the minutes of the *Fun Palace* cybernetics committee meeting, January 27, 1965. Cedric Price fonds, Canadian Centre for Architecture. Courtesy Canadian Centre for Architecture.

be something much more ambitious: a new kind of social organism that would adapt and change based on its inhabitants, bringing them joy and pleasure while monitoring and quantifying them at the same time, much like our experience with sensing machines today.

In other words, the *Fun Palace* reinforces the extraordinary and yet historically grounded idea that immersion and participation could not exist without their counterparts of control and regulation. Contrary to the belief that artists could humanize systems designed for measurement and governance by placing them into artistic contexts, it would be these very processes of sensing, feedback, decision modeling, and prediction that profoundly transformed immersive artistic experiences into a subtle and yet powerful form of social control. Mathematical calculation and modeling designed for war could be transferred into the arts, in a more or less frictionless manner, with aesthetic gratification little disturbed. Indeed, it seemed to be the perfect combination both then and now: social control mixed with sensorial thrill.

⌒

What are we then to make of these two visions almost a half century apart from one another: *Borderless*'s sophisticated sensing machines and the *Fun Palace*'s imagined ones? In comparison to the inventive but unrealized *Fun Palace*, *Borderless* actually exists; it is a materialized thing that you can enter and experience directly with your body at a scale and level of popularity that the *Fun Palace*'s team could have only dreamed of.

At the same time, the *Fun Palace* was after something else: a profound moment of transformation called *transcending the familiar*, achieved through a vision in which politics and aesthetics could meet in a liberatory way. But today, as *Borderless* reveals, control and feedback no longer seem to matter. They are so deeply enmeshed in contemporary artistic experiences that we no longer even think about their once-radical significance to transform not only spaces but also the people that moved through them—to "experiment with new and unforeseen ways to be."[46]

Immersion by whatever means necessary, and not social transformation, is the present and future goal of the new art forms enabled by sensors and computers. In *Borderless*, you are an individual, and your interaction with the environment by way of sensors is almost a distraction rather than a feature. The installation's success comes about due to its status as an artistic thrill ride, the equivalent of a sensory roller coaster. The ability of visitors to influence the space because of their behavior is almost something of an afterthought from teamLab. *Borderless*'s imaginary is ultimately an individual reality—one wholly different from the *Fun Palace*'s collective one.

Marshall McLuhan argued that artists functioned as a "distant early warning system" for technological transformations to come. That the arts would be a warning signal comes as no surprise. Immersive experiences in the twentieth and twenty-first centuries, from visual arts and immersive theater to VR games and teamLab, have more than proven McLuhan's point. These new immersive and participatory experiences have been conceived as environments of shock and wonder that turn the human senses inside out by way of machine automation and sensing infrastructures.

As a sensing machine, *Borderless* perfectly demonstrates that the line between the human senses and the machine sensing environment will become ever blurrier. The sensing machines that constitute the immersive and participatory spaces of contemporary experience culture are no longer just extensions of us. Through sensors, they are their own senses, reinforcing the belief that the machines sense us in order to create a seamless and radical transformation through their own manner of perceiving. But to what ends?

5 Androids That Sing

Nothing is stranger to man than his own image.
—Karel Capek (1920), *R.U.R. (Rossum's Universal Robots)*

The conductor raises the baton and, with a light swing downward, the music commences. The compact orchestra produces slow pizzicato plucks from the strings, with gaps of silence in between. Eventually stronger accents appear together through slash-like bursts emitted by the wind section along with an incessant piano rhythm. The music that gradually pours forth expresses somberness with a glimmer of hope.

Then something extraordinary happens. The conductor, who has been marking time with expressive, periodic motions, swings around toward the audience and begins singing! Weird noises, like the combination of weeping with the first words emitted from a baby's mouth, fill the air. This uncanny, compelling mélange slowly builds for almost five minutes to a climax until, suddenly and without warning, the music and the conductor stop, halting mid-phrase.

This conductor is not your average maestro. They're an android—a mobile robot with a human-like form. The android has been designed by a team around two well-known and idiosyncratic Japanese scientists: roboticist Hiroshi Ishiguro, who famously built a robotic double of himself called a *geminoid*; and physicist Takashi Ikegami, who creates hardware and software machines that exhibit life-like behaviors.

Alter, the name given to the android, is certainly an appropriate moniker, especially in this strange musical performance in which they conduct an orchestra of human players while singing in a foreign language that only they know. The name Alter is meant to suggest both an altered state of consciousness and a machine with an "alternative, non-human mind."[1]

Constructed of pistons and plastic, silicone, metal, hydraulics, and computer code, with a child-like, half-completed face and sensors galore, Alter moves like they are alive. The motions are unpredictable and uncanny. In watching Alter respond to the world, their shifting face, unsettling vocalizations, and sometimes chaotic jerks immediately bring to mind a classic question posed by feminist scholar Donna Haraway in the 1980s of why our machines seem "disturbingly lively" while we as humans appear "frighteningly inert."[2]

Most experimental robots are usually banished to the university research lab, but Alter is different. They are conducting and starring in their own performance, an "android opera" called *Scary Beauty* by the Japanese contemporary composer Keiichiro Shibuya, a long-time collaborator of Takashi Ikegami. Although the opera premiered in Summer 2018 in Tokyo, I have the opportunity to experience Alter in the flesh (so to speak) at a small press conference and performance in the lobby of the New National Theater in Tokyo on a rainy spring day in 2019.

Ishiguro's Intelligent Robotics Laboratory at the University of Osaka has created Alter's mechanical body, but Ikegami's group at the University of Tokyo has designed their "nerves" and "brain."[3] Inspired and shaped by neuroscience, machine learning, biology, and chemistry, Alter's interconnected algorithms drive the robot's physical actions. More startlingly, the android is a new generation of sensing machine because their sensors and

Figure 5.1
Left: Alter the android conducting *Scary Beauty*. Photo by Kenshu Shintsubo. *Right:* Alter's sensors. Courtesy Takashi Ikegami.

algorithms enable Alter to begin to develop a radical property formerly reserved for human beings: autonomy from their creators.[4]

⌒

Takashi Ikegami is the ideal person to design such a machine. A trained physicist who teaches and researches at the University of Tokyo, Ikegami has worked for more than twenty years in an interdisciplinary research field called *artificial life*. In an age of smart machines and artificial intelligence, artificial life, or A-Life, also sounds new but it's a concept that has been around for forty years.[5] Originally named in 1986 by American biologist Christopher Langton, this interdiscipline explores the simulation of living or natural systems and how such systems can exhibit life-like characteristics. A-Life brings together everything from computing, chemistry, biology, and mathematics to robotics and even the arts, architecture, and design. Since the 1990s, hundreds of books and thousands of scientific articles have been written about A-Life processes, including a Japanese book by Ikegami himself.[6]

Throughout its first historical phase in the 1980s, A-Life was primarily concerned with how to simulate life. But this took place mainly on the computer screen. Ikegami's lab is now working on "constructing artificial life in the real world," the next step in A-Life's evolution.[7] Ikegami wants to build artificial systems that can survive in the noise and precarity of the physical world and exhibit autonomy, self-organization, sense making, adaptation, and structural coupling—complex concepts used in A-Life debates that provide the basis for understanding the conditions necessary for something to be seen as living.

To build such systems, Ikegami's lab of graduate and postdoctoral students and visiting scholars conduct research in an interdisciplinary manner that is unusual for scientists. The lab experiments with biology, computers, and machines and publishes in half a dozen disciplines, ranging from computer science to biology, psychology, cognitive science, and neuroscience. Ikegami himself also works as an artist and collaborates with other artists as well while tirelessly engaged in building links between the arts and the sciences, in both scientific and cultural institutions.

I witness Alter performing on this rainy day in Tokyo in 2019 for a specific reason. I've invited Ikegami to collaborate with my own graduate students and me on an artistic commission for a traveling exhibition on the past, present, and future of artificial intelligence called *AI: More Than Human*, to be produced by the Barbican Centre in London, one of the largest cultural centers in the world. We plan to create a fifteen-meter tall light sculpture that will be installed inside the public space of the Barbican Centre, where thousands of passersby walk every day, and we are interested in collaborating with Ikegami and his students to develop new software to control the behavior and actions of thousands of LEDs that make up the artwork.[8] The Barbican project thus catalyzes an ongoing collaboration exploring the nature of building sensing machines and their effects.

Designing a robot that is autonomous is no mean feat. Even the definition of autonomy is hotly debated among artificial life researchers. Normally we think of autonomy from a political point of view: the ability to act on our own, independently and without interference. But in the context of artificial life, autonomy has a different meaning. It loosely suggests two ideas. The first is the potential for what philosophers call *self-governance*—to interact within an environment without continual, external intervention and to achieve or even generate a set of goals. According to artificial life researchers who have studied the creation of autonomous systems, fulfilling these properties is usually the main design goal for robots.[9]

The second idea of autonomy is more conceptually complex but also more interesting, which is why Alter is indeed a special kind of robot. Autonomy means *self-constitution*—the ability of a living entity to produce the mechanisms to recreate its own structure. This idea is captured by the term *autopoiesis*—literally, to *self-make*. Originally proposed by the Chilean biologists Humberto Maturana and Francisco Varela, autopoiesis describes how metabolic (i.e., cellular) systems are able to reproduce not only themselves, but also their mechanisms, the components that enable them to reproduce again and again.[10]

The key to an autopoietic system is that it is an enclosed system. Its behavior is determined by its internal structure and not by external factors. Maturana and Varela say that the system is *operationally closed* to the world. This means that as an entity interacts within its environment by way of its sensors, information from the external world comes into it and perturbs its internal sensory dynamics: nerves, muscles, brain. But this information does not directly tell the entity what to do. There is no one-to-one correspondence. Instead, its actions become what Varela calls *structurally coupled* to the environment. Ikegami gives an example of what such structural coupling entails:

> Structural coupling is different from just having a sensor with information from the outside coming to the inside and then being processed and output and you react. Structural coupling is more like the following. If you go to Shibuya (a very dense and visually overwhelming area of Tokyo) and run around, you don't really know what is going on. You don't know what things are going in and out. You just experience this crazy city landscape. Sometimes you are conscious of what is happening and sometimes you have no idea what you are doing. But all of this is being mixed up and creating impressions and memories. So, it's not a clear boundary between inside and outside. It's more complicated. Sometimes you instantly get information from your sensors but sometimes you need a few hours to get what is happening. You are assimilating to the environment that you are situated in.[11]

Ikegami is after both forms of autonomy for Alter: their ability to sense and interact with the environment without human intervention, and structural coupling to that environment through their own internal dynamics. In other words, *autonomy* is about relations between the world and an entity, not separation and independence. Alter learns to move and behave based on being coupled to their environment over time and without being explicitly programmed to do so.

To artificial life researchers like Ikegami, the term *dynamics* also has a very specific meaning. It signifies the ability of a system to change and adapt over time. As he tells me, dynamics are one of the keys to living systems because we are finite. "Abstract computers like Turing machines have an infinite amount of time to solve problems because they don't have to hurry up. But actual animals have to hurry up because they have finite time. Finite time is the core of living systems. We therefore try to understand what natural computation is in living systems. What time is in these systems and how time is organized and reorganized."[12]

Like many robots, Alter is something of an artificial miracle. Both Ishiguro and Ikegami's research teams have sleeplessly worked for months to give the android an elaborate sensorimotor system. They have constructed Alter's body out of metal and plastic, with the android's half face out of silicone, and enabled Alter to move its arms and head by way of an elaborate computer network that controls motors and hydraulics. Much of this is sophisticated engineering but is still fairly standard for robotics production, particularly coming from Ishiguro, whose geminoid robots have been long featured in popular media worldwide. In Japan, the geminoids serve as tour guides and waiting room attendants and work in Japanese offices, as well as in the only hotel run by robots in the world.[13]

Despite these technical feats, Ikegami claims that these robots are too simple; they exactly fulfill the original meaning of the Czech word *robota*—a worker, forced laborer, or even slave. "There are many of these robots all over Japan. For example, there is a robot from the Denso company who can get beer for you from the fridge. But so what? So simple. I wasn't satisfied with this stupid kind of experiment. Because if I said, 'Can you get the beer for me?' and they said, 'No, I don't want to,' then that would be more interesting. That would be much closer to a living system."

But it is in Alter's sensors and software, the expertise of Ikegami's lab, where the real fireworks happen. Ikegami has given the android audio, visual, and other senses using a complex, interconnected network of sensors: microphones to hear, temperature and humidity sensors to feel the changes in moisture or hot/cold, light sensors to distinguish darkness from brightness, and proximity sensors to know when someone or something moves close to or away from them. These sensors provide the crucial information that allows the android to perceive their surroundings.

Sensors are critical for Alter to gain understanding of the outside world. But they are only the first step toward autonomy. To create the conditions for Alter to have autonomous internal dynamics to process what they sense, Ikegami harnesses mathematical and statistical models that sound like they have emerged from another universe: spiking neural networks, learning models based on neural plasticity and stimulus avoidance, chaotic pattern generators, and reaction-diffusion equations, among others.

That so many different techniques are needed to create artificial life is standard operating procedure for Ikegami. He calls it a *maximalist approach*: "The maximalist approach is like all-in-one ramen in the southern part of Japan. All-in-one ramen is a really crazy thing. It's everything—pork, nori, ginger, whatever. And when you taste it, it's not just a linear combination of stuff. But you can only get something new by putting everything together. It's a response to physicists who want to simplify everything. Maybe we can do this, try something new out—put all sorts of things together in a real situation."

The maximalist approach is clearly evident when Ikegami shows me a complex flow chart of the robot's software. The complicated diagram of interconnected and interwoven boxes, lines, and loops demonstrates that everything folds back on itself, as well as linking up with other components in the system (see figure 5.2, top). Everything seems to influence everything else.

Take Alter's sensors. The data picked up by the different proximity, temperature, humidity, and other sensors doesn't directly tell the robot how to move in response to what it senses. Instead, the data is sent into a mathematical set of equations that creates a new type of dynamics between the different sensors themselves. In other words, Alter's sensor network behaves in an autonomous manner as well. There is no simple input-output model.

Before Alter, Ikegami and his graduate student Maruyama Norihiro had already researched how sensors could be used to try and understand something about *affordances*, the characteristics that help us perceive what is going on in an environment.[14] They wanted "to investigate how an artificial life system would behave in a real, open, ever-changing environment with short- and long-term environmental changes, including human behaviors."[15]

The model for what these scientists call an *autonomous sensor network* came from one of the founders of modern-day computing: mathematician Alan Turing. While Turing is known for his cryptographic work in breaking the code of the Nazi Enigma machine during World War II, as well as for his mathematical model of an ideal computer called the Turing machine, what is less known is Turing's later research in biology—specifically, an area called *morphogenesis* that studies the evolution of form in living systems.

Turing was fascinated with how certain groups of chemicals that he called *morphogens*, which have the ability to react and diffuse together through a medium like tissue, might be a key factor in the development of living cells themselves. He set out to turn this thinking into quantified form, using a set of equations called reaction-diffusion. Reaction-diffusion equations have been used to provide explanations for pattern formation: how certain kinds of patterns (appropriately named *Turing patterns*), such as zebra stripes or the vein structure on a leaf, propagate in living systems.[16]

The basic concept is that two different chemical substances interacting together and mutually influencing each other can produce a stable, repeating pattern.

Ikegami has applied these reaction-diffusion equations to the sensor data in his autonomous sensor network in order to create a kind of artificial chemistry that will influence how the different streams of Alter's sensor data might react to each other. This virtual chemical network influences the sampling rate (i.e., how fast the sensors digitize the incoming sensor data) and shapes how the individual sensor readings mutually interact (see figure 5.2, bottom).

What is most relevant for the development of Alter's autonomy is that all of this conditioning and manipulation of sensor data is done without explicit instructions from the programmers for what to do or how to do it. The reaction-diffusion equations enable the pattern formation between the different streams of sensor data. It's as if the different senses become highly coupled and influence each other in unknown and potentially dramatic ways.

But what then happens to this sensor data? Ikegami has a simple answer to this question, one that forms the real secret to Alter's autonomy. "When several people gather in front of Alter, the distance sensor (for example) receives many inputs. Just as human beings convert such external stimuli into patterns of neurons, so does Alter. The information that is sent from the sensor network then alters Alter's neural cell network."[17]

Alter has a set of autonomous sensors that don't directly feed them sense data but rather different preprocessed combinations and scales of that data. But how does a machine made of metal and electronics have a neural cell network?

What the android hears, sees, and feels is captured by their sensors, digitized and then fed into another piece of mathematical alchemy that creates possibilities for the machine to move and respond to people or events. This mathematical concept, better known as an *artificial neural network*, is more than a century old. Ikegami uses artificial neural nets for Alter as well, but he takes a different approach than most computer scientists. To understand this requires us to take a brief historical detour to understand the basic principles behind Alter's "mind."

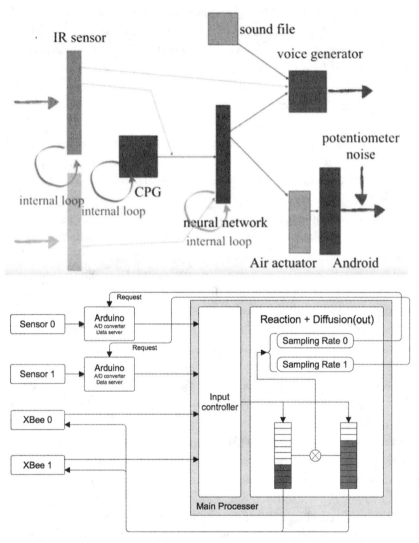

Figure 5.2
Top: Alter's senses. *Bottom:* Diagram of autonomous sensor network. Courtesy Takashi Ikegami.

Neural networks are computational entities that run our current world. They read our checks, decide on bank loans, play our music, send people to prison, and analyze our body scans. They underlay Google's search engines, Facebook's news feeds, Netflix's movie recommendations, and Spotify's preference models, enabling these commercial companies' video and audio systems to "know us" in order to offer ever-new customized content. There are also growing concerns that certain varieties of these networks (but not all) don't play fair—that there is inherent racial and gender bias directly baked into both the structure of the algorithms and, more importantly, the data sets that train them.[18]

Despite its name, however, an artificial neural network is not a biological entity. It is a mathematical explanation of a biological process, a vastly simplified model of how neurons in the brain might function. Unlike real neurons that send electrical and chemical signals between other neurons, artificial neurons send out only numbers.[19]

The concept already appears as early as 1903 in the work of psychologist Sigmund Freud, who attempted to explain the flow of "nervous energy" circulating through what he called *networks of neurons* inside the brain.[20] But the mathematical proof had to wait for forty years later, when such neurons were quantitatively described in an influential 1943 scientific paper by psychologist Warren McCulloch and logician Walter Pitts. McCulloch and Pitts claimed that neurons had an all or nothing quality: either they fire, producing a "spike" or "action potential" that sends electricity and chemicals to other neurons, or they don't.

A biological neuron essentially does three things. It gets an input signal from another neuron (called a *dendrite*), processes that signal in its cellular structure (the *soma*), and, based on some set of conditions, sends the output to other neurons (*axons*) when it spikes or fires. The link or connection between the neurons is what is called a *synapse*.

These three principals were therefore enough for McCulloch and Pitts to reimagine the neuron not as a flesh-and-blood biological entity but as a little computer: a "logical machine." McCulloch, a trained psychologist and psychiatrist, had long been searching for an internal logic of the nervous system, and he found it by rethinking the biological neuron as an abstract computer that could produce a binary output: 0 or 1, yes or no, true or false.

Because these so-called Boolean operators were considered the foundation of contemporary logic, McCulloch and Pitts took the next step by claiming a single neuron could produce a logical proposition—a true or false statement.[21] Neural signals could thus be shown to be equivalent to logical propositions. Today, in digital circuits, this concept goes by the name *logic gate*—a simple component that takes two inputs, compares them using a set of logical operators like AND, OR, or NOT, and then outputs a signal to the next components in the network, depending on what it finds.

McCulloch and Pitts claimed that if these logical calculating machines were connected together in a net of neurons, they could be used to link different logical propositions together to solve more complex problems. The title of their famous paper, "A Logical Calculus of the Ideas Immanent in Nervous Activity," which suggested that neurons could indeed hold and calculate concepts or ideas, quickly gave their game away.

Interestingly, a mere nine years after McCulloch and Pitts's model, another early formulation of neural networks was published, this time from a more unusual source. The Austrian political philosopher and later Nobel Prize–winning economist Friedrich A. Hayek wrote an unconventional book (for an economist) called *The Sensory Order: An Inquiry into the Foundations of Theoretical Psychology*.[22] In the book, Hayek sought to understand a question that had long vexed neuroscientists and psychologists: What is the relationship between the world outside of the mind and the world inside?[23]

Specifically, Hayek wanted to know how the mind determines the distinction between different sensory qualities—in other words, how does our nervous system distinguish one kind of stimulus (e.g., sound) from another

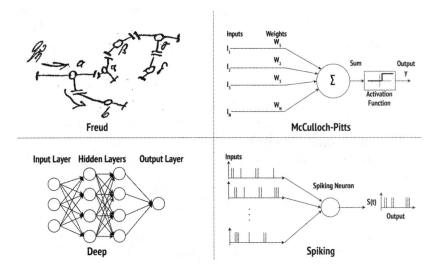

Figure 5.3
Different neural networks.

(e.g., light) so that we have different experiences? To address this question, Hayek exhaustively set out to trace the very structure of our sensations, from their origins in physical stimuli in the world through impulses in the neurons and, ultimately, the brain itself. He proposed what he called the *sensory order* and claimed that this sensory order was made up of three layers: (1) the physical order of the external world and its physical stimuli; (2) the neural order of fibers and the different impulses or electrical spikes that occur in those fibers; and (3) the mental, phenomenal, or sensory order of mental thoughts, images, and sensations.[24] Hayek's central aim was therefore to demonstrate how these three orders operate on each other.

Hayek named the ability to differentiate between one sensation and another *classification*, which denotes the neural system's ability to categorize one thing or object from another. In this way, perception is "always an interpretation, the placing of something into one or several classes of objects."[25] We can understand the whole of perception as an ordering system.

Where does this process of classifying stimuli into different categories that constitute the sensory order actually take place? For Hayek, it happens in the organization of the neural system, through the "system of connexions by which the impulses can be transmitted from neuron to neuron."[26] Different configurations of neurons classify different sensory stimuli, and thus the sensory order, all of our mental or phenomenal experience, is based on this system of changes of connections between different networks of neurons.

But this sorting out of different stimuli according to patterns of different neural connections is only the first element in Hayek's theory of perception. The other crucial part is memory. Our experience of events in the environment is formed by what Hayek calls *past linkages*. Sensory experience is dependent on a store of *accumulated knowledge*, an "acquired order of the sensory impulses based on their past co-occurrence." Radically, sensory experience is not a product of the mind but vice versa: the mind, the ever-changing reconfiguration of neural connections, is a *product of our experience*. In other words, sensing comes before mind.

The fact that our sensory experience in the present is based on previous experience creates a conundrum because a portion of what we know is actually not learned from our direct experience but is the result of past experience. Our experience is "determined by the order of the apparatus of classification which has been built up by pre-sensory linkages."[27]

This statement is not to be taken lightly, not only for Hayek's dynamic conception of the sensory order but also for the much broader question of how we come to know what we know through our human senses. For the fact that our experience is path-dependent—based on prior experience—illustrates that not all sensation is, in fact, even open to our direct experience. Like other complex orders, such as the spontaneous emergence of patterns in flocking birds or the intricate behavior of financial markets, the mind's dynamic relations and interactions cannot be directly experienced by us.[28] There is, as Hayek phrases it, "a limit to explanation" in terms of what we can know; a complexity and order of self-organization that simply eludes our ability to grasp its overall operations.[29]

This makes sense if we think about the structure of how our brain works. We know that somehow the firing of billions of interconnected neurons is linked to our senses of hearing, seeing, touching, smelling, and tasting, but we have no access to how this neural order and its dynamics leads to the higher-level mental states that we call *sensory experience*. The same might be said for the complex infrastructure of sensing machines as well.

Importantly, Hayek's description of the mind as a classifying machine, its bottom-up, spontaneous, and self-organizing structure, anticipates the paradigm of neural networks by two decades, specifically, the idea of *connectionism*. Connectionism, which also goes by the name *parallel distributed processing*, describes a model of mind and intelligence that "emerges from the interaction of large numbers of central processing units."[30] This model forms the basis for the modern concept of a neural network.

Intelligence and sense making in the connectionist paradigm don't begin with abstract rules or symbols. They start with simple computational units that are activated and inhibited by other like units and, "when appropriately connected, have interesting global properties that embody and express the cognitive capacities being sought."[31] Intelligence emerges based on the system's connectivity, "which becomes inseparable from its history of transformation." To put this in Hayekian language, the mind spontaneously emerges based on previous linkages and experiences that have been encoded in the network of neurons.

Neural networks have advanced in computational power and by orders of magnitude of complexity since the 1940s—but in many ways, their basic properties remain the same as McCulloch and Pitts and Hayek imagined them: mathematical units that do computations and send those numbers to other interconnected neurons based on the strength of connections between them. If a certain set of conditions is met by what is called a *threshold* or *activation function*, the neuron *fires*; that is, it sends its numbers to the next neurons connected to it.

In contrast to the McCulloch and Pitts model and Hayek's more theoretical ideas, in current deep learning networks, different groups of neurons handle different characteristics or features in multiple layers: an input layer, a series of hidden layers, and an output layer. The hidden layers are made up of a number of interconnected neurons. When patterns such as images or time-based signals like speech are presented to the network via the input layer, the signals are output to one or more of the hidden layers, where the actual processing is done via a system of connections that resemble synapses. After processing this data, the hidden layers then send their numbers to a final output layer, where the result of the network's calculations is output.

Most importantly, many neural networks can adapt over time through what is called a *learning rule*. They can adjust the strength of their connections or weights between different neurons according to the input patterns the neurons receive, a concept proposed by Canadian psychologist Donald Hebb in 1949. Hebbian learning mathematically describes that the connections between two neurons might be strengthened if the neurons fire simultaneously—invoking the expression, "Neurons that fire together, wire together." The rule states how much the weight of the connection between two neurons should be increased or decreased in proportion to the product of their activation and hence is a key idea in how neurons learn different kinds of input patterns over time. In later works, Hebb himself even acknowledges Hayek, and the late Nobel Prize–winning neuroscientist Gerald Edelman does the same for both.[32]

Surprisingly, in a discussion with Ikegami, Hayek's name comes up. It seems that this physicist and complex systems scientist has read Hayek's *The Sensory Order*. Not only that, Ikegami has been inspired by Hayek's description of the three layers of the sensory order: the world of stimuli outside the brain, the organization of neural firings inside the brain, and the production of mental concepts or images as a result of those firings. In this sense, Alter's brain and nerves are also made up of artificial neural networks. But these nets are not like that of McCulloch and Pitts or the deep learning networks that power Google or Facebook. Inspired by Hayek's concept and structure of the sensory order, Ikegami's team has written software to use what are called *spiking neural networks*. Spiking neurons are also not biological. Like McCulloch and Pitts's little logical computing machines, they are mathematical representations; that is, simulations. But spiking neurons also have another key difference from other artificial neural networks. *They produce time.*

How do mathematical models of neurons produce time? Spiking neurons send out sequences of numbers that represent discrete bursts that suddenly can rise, peak, and then resettle back to their initial value before they perform the same action a few milliseconds later. These bursts/spikes are not isolated in time but come in sophisticated patterns and sequences. Not only is the spiking model conceptually different from other artificial neural networks, it is mathematically different as well.

Named after the Russian mathematician who first described it, the Izhikevich spiking model that Ikegami uses is based on a single differential equation—a mathematical representation of how a continuously varying quantity also changes over time.[33] The equation basically models how the electrical potential of a biological neuron's structure, its membrane, changes over time before it sends out a spike. Biological neurons fire only when an electrical and chemical threshold within the neuron is exceeded, and Izhikevich's differential equation renders these complex physiological dynamics in relatively simple mathematical form.

But why go to such effort in using these spiking neurons over the neural networks that have efficiently proven that they can recognize everything from images of cats to the sound of an Italian voice? Ikegami has a clear answer to this. "Logic doesn't help time. I'm interested in time since this is the key to a living system. Time is when you have spiking patterns in the brain. Timing is a critical issue in spiking neurons and is necessary if you want to understand the brain as a time-generating machine."

What then is the link between Alter's sensors and their spiking neurons? In biological systems, sensory stimuli trigger different sensory receptors in the sense organs to fire in different spiking patterns that are then sent to the brain to be further processed. Alter's "altered" sense data from their autonomous sensor network also sends this data to their mathematical spiking neurons, and the data changes the strength of connections between the neurons. This is one way that Alter learns to respond to things in their environment.

Like all of the android's components, however, the sensory data isn't simply routed directly into the spiking neuron model. Instead, Ikegami has also built another piece of software to mimic biological systems: a set of seven central pattern generators or oscillators that create unpredictable rhythmic behaviors and that influence the android's motor movements. These seven oscillators receive spike patterns and then send commands to the different motors in their head and arms. The resulting movements of Alter are not rigid and mechanical; they are oscillatory and continually in motion.

"Robot people usually use stable oscillators. We use oscillators that behave chaotically; this means that their periodicity becomes quasi-periodic—not always predictable. I think this is a very natural way to see living systems as consisting of many cycles and feedbacks. I was interested in understanding a body not as stable but as a coupling of chaotic processes." In other words, the pattern generators spontaneously create different rhythms, and the spiking neurons produce variations on these rhythms by perturbing the pattern generator, which controls the different motors. Another set of loops within feedback loops. In this way, Alter is also driven by external information that "makes it possible to simultaneously create complicated movements and respond to the environment."[34]

Ikegami gives a compelling example of why such processes are necessary for Alter as a music conductor. In the complex dynamics that make up an orchestra, conductors are assumed to be timekeepers—to keep a stable, periodic rhythm that organizes the players:

> So we have the android conducting the orchestra. If you put a metronome in front of an orchestra, it's very stable, but it doesn't mean that the orchestra becomes fantastic as players. The metronome is not good enough to bring the orchestra together. Maybe the metronome instead needs to couple with the orchestra. What kind of coupling? Maybe the robot cannot be a stable metronome; it's a more adaptive metronome, resonating with the orchestra. That's why the movements of Alter and their periodicity are not a perfect metronome. The orchestra and the robot have to be coupling with each other, resonating together.

Alter's complex network of sensors, spiking neurons, and pattern generators provide the technical conditions for the android to develop their own internal and autonomous dynamics that interact with the surrounding world. As Ikegami states, "The sense data goes through these neural networks and the neurons have their own autonomous activity and memory. This is why Alter is more lively, reacting to people. In fact, those who interact with Alter cannot easily predict what the android will do."

Alter's behavior might be said to engage in a more crucial property necessary for autonomy—what is called *sense making*. Sense making denotes making sense of the world through a structural coupling between the robot and its environment. In other words, sense making is an active participation in the world and not passively waiting for data to pour in. This is why sensing is so critical for Alter's proposed autonomy. In the early versions of the android (there have been new versions), the sensing was still rather crude. As Ikegami states, "To capture the complexity of the external environment, we must introduce richer sensors, since greater sensations produce higher cognition."

Ikegami argues that behind each sensor is a mind. Like Hayek's argument that mind doesn't precede experience, Ikegami also claims that mind doesn't come before sense but rather emerges out of different sensors and the environment those sensors sense. "If you change the kinds of sensors, different kinds of minds will emerge. We have to be careful in terms of which kinds of sensors we choose—to see what kind of mind will self-organize."[35]

Self-organization is another cornerstone of artificial life. It is the ability of a system to achieve structure and pattern due to the collective interaction of the individual elements in it itself rather than from a central, exterior authority. Researchers call this the *bottom-up approach*. Through self-organization, we then might catch a glimpse of the true significance of Alter as a genuinely new kind of sensing machine.

Since Alter's *Scary Beauty* debut in 2018, the android has been busy, with starring roles in international exhibitions and performances from robot festivals in Germany and the *AI: More Than Human* show at the Barbican Centre in London for an audience of over one hundred thousand visitors in the summer of 2019 to a Canadian film about the fear of dancing.[36] It makes sense that what Ikegami calls a "new species" gets their first public exposure and performances mainly in the field of culture and the arts. As we saw from the histories of immersive and participatory environments, the arts are a safe place where the imaginaries of scientists and artists can have free reign.

But Alter also demonstrates that an artificial entity whose senses constitute an interlocked combination of electrical, mechanical, and software components can learn something about the inherent spatial and temporal characteristics, the affordances, of the environment in order to shape their perception and action within that environment.[37] A robot gains autonomy by coupling with the environment based on these sensors. As robot and environment interact with each other, we can therefore say that Alter brings forth or *enacts* a world based on their own position and sense-making processes as part of that world. According to Ikegami, the different spatial-temporal actions that give an entity like Alter their particular dynamics thus transforms the kind of mind and, hence, experience that may emerge in this quasi-living thing.

III Engineering

6 Sensing on Wheels

It'll be an order of magnitude safer than a person. In fact, in the distant future, I think [. . .] people may outlaw driving cars, because it's too dangerous. You can't have a person driving a two-ton death machine.
—Elon Musk, Tesla founder

In 1986, a little-known German aerospace engineer developed a sensing machine with far-reaching consequences. Ernst Dickmanns was working at the German Bundeswehr University, a research institution founded by the *Bundeswehr*, the German armed forces, that educated military students and officers. Having grown tired of his earlier research designing systems that calculated the trajectory of space vehicles reentering Earth's atmosphere, Dickmanns turned his engineering prowess toward a more intriguing moving object: the automobile.

Dickmanns's new research program wouldn't remain in the controlled environment of the engineering lab for long. It was ultimately destined for the wild and noisy real world. With miniscule funds, he soon purchased and outfitted a five-ton Mercedes Benz van with two cameras set on a motion rig containing accelerometers and inertial sensors, together with an onboard computer and actuators (motors) that would control the steering, acceleration, and braking of the vehicle. The hulking van's awkward identifier, VaMoRs, an acronym for the German *Versuchsfahrzeug für autonome Mobilität und Rechnersehen* (Experimental Vehicle for Autonomous Mobility and Computer Vision), would unknowingly name a vision of things to come: the self-driving car.

Before this work, Dickmanns began in a different place. He first had become intrigued by the possibilities of teaching computers to see, but there was a problem. In the 1980s, computers were analyzing static, 2D images, which was a process that ran at a snail's pace and made it near impossible to use in real-world situations in which the size and shape of objects were in three dimensions (3D) and changed over time, depending on the perspective of the camera. The seeing process, which took on average ten to twenty seconds, was achingly slow because researchers analyzed only a single image or a small sequence of images at a time to look for many *features*: the edges, corners, and general visual characteristics that constitute an object to a machine.

Dickmanns's understanding of how humans saw, especially when they were in motion, drove his particular approach to machine vision. The human eye doesn't continually process the entire image that hits the retina. Instead, it directs most interest to those features that change in the visual field. In other words, what is important is not only the content of an image "but the relatively slow development of motion and of action over time."[1] The engineer and his graduate students took this fundamental principle of human vision as the basis for using cameras and commercially available computers to analyze the position and motion of objects in three-dimensional space as they changed—what Dickmanns called the 4D approach (three dimensions plus time).

During the early phases of their research, however, the team ran into a critical obstacle. They realized that there was no intrinsic link between the 2D image the camera saw of the world "out there" and the internal representation of that world inside the programming code. To address the issue, the engineers thus created an internal world, a complete 3D model of the real, external world, inside the machine. They then used a series of mathematical and statistical techniques to predict how specific features of objects in the model might change over time. The 3D world in the machine would be compared constantly to the real world captured by the camera, and the prediction of where objects might be in the model would be updated based on where they actually were as time passed. This ever-changing understanding of seeing based on the progression of time suggested to Dickmanns the ideal name for his system: *dynamic vision*.[2]

While experimenting with his five-ton van, Dickmanns also realized something else. The sense of sight obtained through the camera was not enough;

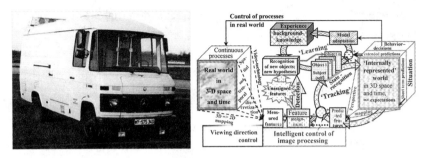

Figure 6.1
Left: Ernst Dickmanns's Mercedes van. *Right:* Schematic of dynamic vision.

it was too slow. Because images contain so much data, trying to recognize things in them, even simple forms like the edges and corners that make up more complex objects, introduces lags on the order of a few seconds—what engineers call a *time delay*.

Time delays happen in biological systems too. We don't take in the entire world with only our eyes. Seeing is linked or correlated with other senses, such as balance, awareness of one's position in space, or the sense of hearing the surrounding environment. All of these different sensory systems have different time factors in how they process data, with some being faster than others. To compensate for the lag in one sensory system, the nervous system thus integrates or fuses the rich data coming in from the other senses at different spatial and temporal rates in order to get a total picture of the world.

Machines have no such capabilities. While they may have dozens or even hundreds of sensors, they don't know they have these senses. Engineers therefore devised a trick—a technique called *sensor fusion* that allows them to aggregate and combine multiple feeds of sensor data together in order to gain an accurate portrait of the sensor field.

For Dickmanns's self-driving van, sensor fusion had a central goal: the reduction of uncertainty in the real world. Sensors ultimately are designed to deliver measurements of things: temperature, acceleration, vibration, or the size and position of an object in 3D space. Just as uncertainty in the human perceptual system arises when we suddenly can't see in the dark or when one kind of object blocks or occludes another, uncertainty in a sensor occurs if it malfunctions, has limited spatial or temporal coverage, or produces inaccurate, ambiguous, or noisy (random) measurements. The major premise in sensor fusion then is the realization that more sensors are better than one.

But there is a trade-off. No single sensor can accurately give us a measure of the total environment, so we must try and predict (i.e., guess) how the whole is greater than the individual sensors themselves. The statistical methods with names like Kalman filter or Bayesian estimation used in sensor fusion attempt to remove this uncertainty by taking educated guesses and then calculating what the potential predicted outcome might be.[3]

To have a machine drive a vehicle, that "two-ton death machine" that Elon Musk speaks of, seeing would not be the only sense to be deployed. Dickmanns made use of many sensors and fused their data to understand the complex picture of what would take place in the here and now of a five-ton machine barreling down the German autobahn at 96 km/h. Sensing speed from the odometer, acceleration, inertia (the shifting of gravity), tilt and orientation, sound, and even the vibration of the camera were necessary to get as much data as possible so that the system could react to the ever-changing conditions encountered in the harrowing act of driving. In contrast to computer vision, dynamic vision was not only about seeing. It was about sensing and acting on the world.

To create a true sensing machine, Dickmanns also endowed his autonomous car with a capability that was unusual for an artificial device: perception of the environment. When it comes to machines, *perception* has a specific engineering-based meaning—measuring the state of the environment through sensors and then estimating that state in order to decide how to act on the information. Philosophers of mind and neuroscientists call this process the *perception-action cycle*: "the circular flow of information that takes place between an organism and its environment in the course of a sensory-guided sequence of actions towards a goal."[4] Each action causes a change in the environment, which is processed by the organism's senses. This then leads to an action by way of the organism's motor or effector system. These new actions cause new changes that are again picked up the senses, are analyzed, and lead to a new action. The loop then continues to repeat itself.

In essence, Dickmanns's vehicle's ability to perceive its environment gave the machine essentially limited cognitive capabilities. Indeed, before Google, Waymo, Uber, and the DARPA Grand Challenge, Dickmann's early research toward turning the car into a sensing machine had revealed an important fact.[5] Because it needed to operate in the real world, the car suggested a step up by an order of complexity. In contrast, the accelerometers in musical instruments or game controllers, the cameras in art installations, or even the vision sensors of the Kinect are all relatively simple sensing systems. These sensors do not perceive in any holistic way the environment beyond the human user who wields a device or moves their body. But the self driving car raises the ante. It has a clear goal (driving), must navigate a much more complex set of spatial and temporal conditions in the here and now, and, most importantly, makes predictions of what it will do next. For perceiving and acting on the world, one sensor would never be enough.[6]

⌒

Some thirty-four years later, Dickmanns's vision of a mobile sensing machine is on test tracks and even roving some neighborhood streets around the world. Perhaps more than any other industrially produced consumer good, sensing has thoroughly transformed not only the automobile but also the experience of driving itself. If BMW tags its cars with the line "the ultimate driving machine," they might soon be rebranded as the ultimate sensing machine.

In 1970, there were virtually no sensors in automobiles. By the late 1980s, cars were the second-largest industrial market for sensor technologies.[7] New cars rolling off the assembly line are engineered with sixty to one hundred standard sensors, with economy cars having the least and luxury cars the most, demonstrating that automation is never far from your socioeconomic class.[8] An ever-growing multibillion dollar market, the automotive sensing industry forecasts some two hundred different sensors per vehicle by the end of the 2020s. Multiply this number by the number of cars produced each year (twenty-two million) and we quickly reach an astronomical figure: a total of approximately four billion four hundred million sensors installed in new cars in the year 2020 alone.

Sensors in cars are not only designed for augmenting or, if Silicon Valley utopian visions have their way, replacing human drivers in order to deal with the complexities of the road. They also handle almost every kind of mundane task associated with a car as well, from clocking speed and monitoring tire pressure to measuring exhaust and managing steering and braking. Aside from obvious things like video-assisted backing up or side view mirrors that blink, alerting you to the proximity of another vehicle when passing, car sensing is still mostly invisible to us.

But there is another emerging area in automobile sensing that affects not just how a car monitors its own internal processes but how it brings us directly into its perception-action feedback loop in a tangibly felt way. These sensing systems envisage a more radical transformation of human perception—making those of us who drive a key element of the sensing machine by enhancing our perceptual, emotional, and bodily experience while piloting the car.

To accomplish this new form of extreme interaction between the car and our bodies, we only have to explore some intriguing research projects in order to shed light on what might be coming. Automobiles have become veritable sensory laboratories on wheels. Economy and luxury cars alike are increasingly focused on how to trigger sensorial experiences in the buyer. With everything from the quality of materials in the seats to real-time acoustic adjustment of audio and even newly emerging scent systems that affect your mood, almost every aspect of buying a car is now based on "what you feel in your stomach."[9]

Inevitably enriching the automobile industry's bottom line but not reducing the ultimately pernicious environmental effects of driving, the rush toward using sensing to improve the experience of being cocooned in the total artificial environment of the car still makes some experiential sense. With the mind-boggling statistic that Americans alone spent seventy billion hours in cars in 2019, the need to transform this daily routine into a new kind of sensory odyssey seems paramount to ensuring piece of mind while mitigating the ongoing scourge of road fatalities.[10]

Three core areas pointedly demonstrate how reenvisioning the car as a sensing machine has radically transformed the emotional and bodily experience of driving: safety, mood enhancement, and stress management. Safety seems to be one of the main forces leading this transformation. According to the World Health Organization (WHO), road accidents kill 1.35 million people worldwide per year, with 93 percent of these fatalities occurring on roads in low- and middle-income countries, even though such countries have approximately 60 percent of the world's vehicles.[11]

Fatigue and falling asleep at the wheel are the chief causes of car accidents worldwide. To help alleviate this situation, car safety therefore marks an essential domain of research for the incorporation of drivers' bodies into the sensing-feedback-action-perception loop.

These safety systems are already operating in the background without our knowledge. Standard sensors like cameras and scanning radar are constantly detecting and correcting driver-caused anomalies such as drifting out of the correct lane or sudden increases in a car's acceleration. More exotic systems, from Daimler's Attention Assist to Volvo's Driver Alert Control stand ready with steering wheel–based cameras, accelerometers, and microphones to capture a driver's closing eyes or slumping, drowsy head.

Such complex combinations of sensors, electronics, actuators, and microprocessors immediately notify drivers who might fall asleep by triggering wake-up stimuli: alarm sounds, buzzes, beeps, blasts of cold or hot air, and scents. One IBM research study even suggests the implementation of an AI-based "artificial passenger," which involves conversation between the driver and the artificial passenger, corny jokes, or (if all else fails) a jet of cold water straight from the dashboard.[12] Research is also investigating how to use the sense of touch in order to startle drivers out of drowsy sleep before calamity occurs. Intense haptic and vibrotactile stimuli in the form of gentle vibration alerts, artificial massages, low-frequency "blows" to the body, and even vibrational patterns in worn haptic jackets that impart touch-based stimuli are all being explored as solutions for those who might suddenly violate the rules of the road.

One of the most curious research projects gathering preliminary interest from Toyota comes from a Japanese researcher who designed a wearable device using a rather bizarre technology: what is called *galvanic vestibular response* (GVS). GVS is a technique emerging from late nineteenth-century research in which small amounts of electrical current are directly applied to the mastoid bones behind the ears, thus disturbing the sense of equilibrium that is normally regulated by the vestibular system—the inner ear sensory system that provides information to the brain about our balance, position, and orientation.

Using the GVS principle, Keio University researcher Masahiko Inami proposed a kind of vestibular augmentation, exploring how electrical stimuli of a particular nerve called the *vestibulocochlear nerve*, which is partly responsible for hearing and for our sense of balance, can induce the feeling of disequilibrium and acceleration in subjects. The title of the publication that reported on this research clearly signals the earth-shattering sense that car and insurance executives see in the device: "Shaking the World: Galvanic Vestibular Stimulation as a Novel Sensation Interface."[13] Demo videos that still circulate on the internet show a subject walking on a straight line, like undergoing a drunk driving test. Suddenly, these bodies veer off course, twisting at weird angles while a researcher almost outside the frame pushes a button on a remote-controlled device.

Why would an auto manufacturer be interested in a technology that throws one off balance and potentially induces motion sickness? After all, these are not the best physiological sensations to experience while driving a car. In fact, with driver simulation training systems in the laboratory, engineers sometimes attempt to artificially induce the sense of acceleration or vestibular confusion that we might experience during a car accident or even during normal driving when experiments are conducted. But the implementation of such technology during the actual physical act of driving itself, as direct bodily stimulation to suddenly startle a driver who has carelessly fallen asleep at the wheel, takes the next step.

Manufacturers are also hurrying to close the perception-action loop between driver and car with sophisticated automated sensor systems that can either take over in the midst of a collision or warn the driver ahead of time that such an accident is about to take place. In what are called advanced driver assistant systems (ADAS), automated configurations between a human and a car/machine augment or fully assume a human driver's control if they come close to a critical situation. Such systems are already in use for noncritical situations like parking or backing up.

Similar to bats' direction finding through echolocation, the cars use ultrasound to bounce high-frequency sound waves off of potential obstacles behind or in front of vehicles to alert the driver of their presence. Similarly, automatic-braking systems triggered by using wheel speed and rotation sensors to detect speed and torque, as well as radar, also reduce the possibility of rear-ending someone. Technologies that are now being implemented in Toyotas and Fords, among other brands, use infrared and stereo camera vision to take control of the gas pedal directly from drivers if they detect the appearance of sudden obstacles, to slow the car down and reduce acceleration-based impact.[14]

These automated sensing control systems still focus on what researchers mysteriously call the *ego vehicle*—the car as a singular, first-person entity isolated from other vehicles. But research programs waiting in the wings are reimagining the ego vehicle as well. They see the car as only one element in a much vaster network of other sensing machines. The fantasy of the car not only as a collection of sensors but as a sensor among other sensors led one former BMW senior VP to announce that the automobile would no longer be considered a driving machine but instead a "media center."[15]

Research and development in the European Union and Korea, among other places, is advancing work on so-called smart car sensor networks (SCSNs) that could be used for more accurate road environment monitoring. The principle behind such networks is to reenvision the car as a sensor communicating with other cars—to use a car's existing sensors to transform it into a "diffuse network of road-environment monitoring nodes."[16]

A group of Italian researchers, for example, is proposing to use standard *intervehicle* car sensing systems, like accelerometers, gyroscopes, temperature, humidity, and pressure devices, as implicit *intravehicle* (between vehicle) sensors to communicate with other cars about their state. Such implicit sensing means that no new sensors need to be installed. Instead, the systems draw on data from already existing sensors, from the somewhat ordinary (monitoring outside temperature) to the more unusual, like sensing when the auxiliary lights are turned on in order to know when a car drives into a bank of fog.

In a sudden blanket of fog, for instance, one car could alert the driver of another car down the road of the presence of fog so that driver can react before they drive into the mist and have a collision with unseen vehicles. In relation to car culture, such sensor networks are also gaining ground in a more humdrum context that, at first, doesn't sound particularly life threatening: alerting drivers of empty parking spots in congested cities. One central site of deployment for such an application occurred in San Francisco, a city denoted by the *New York Times* as a "modern hellscape" due to its ever-increasing gap between tech riches on the one hand and widening poverty and homelessness on the other.[17] After a nineteen-year-old was stabbed to death in a fight over a parking space in an incident in 2008, the city installed scores of wireless sensors in 8,200 parking spaces to alert drivers of empty places in real time.[18]

Clearly, safety is a core factor in transforming cars into modern-day sensing machines. But another area, called *mood enhancement*, brings drivers into the machine perception-action feedback loop in somewhat subtler but no less marketed ways. One aim for car sensing technologies has long been enhancing our acoustic experience while driving.[19] The need to shape the car's acoustic environment in order to reduce noise both inside and outside as a way to heighten driver alertness and pleasure has become a major preoccupation of the car industry, and sensors help lead the way.

Take the standard car stereo. The fabled and expensive German car radio manufacturer Blaupunkt conceived of real-time acoustic control of the car's interior back in the 1980s with its high-end Berlin stereo tape deck model, a device that certainly didn't shy away from announcing its elite status for

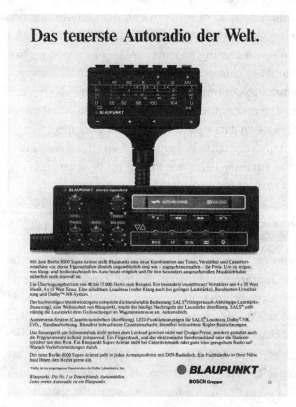

Figure 6.2
Blaupunkt ad, circa 1983.

the then 1 percent. The Berlin was delivered standard in the Carrera 911, German manufacturer Porsche's flagship sports car, and was advertised in Germany as "Das teuerste Autoradio der Welt" (the most expensive car radio in the world), together with the slogan, "Some day most drivers will enjoy car stereo of this sophistication. But for now only Blaupunkt owners do."[20]

The Berlin's minimalist design was something of a fetish object for auto audiophiles. In addition to the in-dash radio/tape deck, the Berlin featured a futuristic tuner and volume control mounted on a flexible gooseneck stalk that could be oriented toward the driver or passenger and worked by touch. Most importantly, this top-of-the-line music-playing device already realized the automated vision of acoustic car interior bliss: the Berlin could raise and lower its volume based on a sound ambient level sensor (SALS), a small microphone that monitored the changing acoustic conditions of the car interior.[21]

This unusual notion of sensors helping turn a car into an ideal listening environment quickly gained speed among other manufacturers. The Bose loudspeaker corporation was one of the first to pioneer the use of psychoacoustic principals in the design of custom car audio environments. *Psychoacoustics* is the science of how we hear, using methods derived from psychophysics, perceptual psychology, and acoustics.[22]

One of the central ways to shape the acoustic environment of a car is to take a cue from psychoacoustics in altering what is called the *frequency spectrum*. The spectrum consists of the range of frequencies and amplitudes (the strength or volume of a frequency) contained within a signal—in this case, audio. Using a *dummy head*, a conventional testing device that is essentially a mannequin with two microphones embedded within its ears, Bose engineers measured different frequency responses at different sitting positions in a car in an effort to design so-called balanced frequency equalization for each seating location. In other words, the audio system would sound good from all seating positions.[23]

Today, real-time spectral analysis, measuring the ever-changing frequencies in the acoustic interior of the car in real time, is a standard feature in most mid-sized vehicles and nearly all luxury cars, with immersive audio systems featuring multiple loudspeakers, automated equalization, and real-time noise cancellation. For example, in 2019, Bose introduced its new QuietComfort sensor-activated noise cancellation system, which uses accelerometers to measure the degree of unwanted vibration on the car's tires and axels to "electronically control unwanted sound." These accelerometer signals are calculated in real time by a proprietary frequency analysis algorithm that immediately generates counternoise that is then played into the car's interior.[24]

Audio control technologies can also create wholly fictitious sounds. BMW uses its audio system to play samples of different engine sounds that match the current rpm and speed level of the moving vehicle in order to fool drivers into thinking that the engine's motor is bigger than it is. Porsche has advanced this trickery. It created an electromechanical noise-inducing system called the *sound symposer* to give drivers a deeper, acoustically emotional attachment to engine sounds. The sound symposer technology monitors intake vibration between the throttle valve and the air filter of the car through the use of a tube-like *Helmholtz resonator*—a device that consists of a hollow container that amplifies the particular frequency of an object as it gets close to the vessel. In the Porsche system, with the push of a button that opens a valve, the amplified, resonating engine sounds are directly transmitted into the interior of the car! The result is a new kind of acoustic relationship between the driver and their car—"a direct acoustic link."[25]

Modifying existing sounds or introducing wholly new ones into a car's interior to shape emotional response is only one strategy in a much larger project to design sensing systems that can monitor and even shift the emotional state of drivers. Sensing a driver's *mood*, defined as a state or quality of feeling over time, has long been the desire of auto manufacturers anxious to know something about their customers while they are directly engaged in the act of driving.

As early as the mid-2000s, the auto press was awash with announcements that Toyota in concert with Sony was experimenting with a "mood-sensing" car—a car that could become a direct "extension" of its driver's feelings. What was revealed was something other than expected. Toyota's patent in fact describes a "device for detecting condition of an occupant of the vehicle" and technologies to transfer how the driver feels directly into the "expression" of the car.[26] How the system does this is an enigma, but it definitely fits into the typical stereotype of the Japanese culture of *kawaii*, or cuteness.

Sensors would operate to detect the car's interval, direction angle, and speed. This data would then be converted directly into anthropomorphic expressions, displayed on the car's exterior. The ability to vaguely express the emotional states of a car's driver is imagined to go far beyond the limited means of a horn or a turn signal. The car instead could be seen as an animal or human body. The headlights, antenna, windshield, and exterior panels would be "the vehicle's eyes, tail, and a body surface" and the car's emotions expressed by changing the color of the exterior panels—making the antenna limp or straight, or shifting the position of "eyebrows" inset above the headlights.[27] In an age when the media is filled with daily reports of sensors tracking our every move, Toyota's early vision seems almost innocent. Its inventors even claimed without the slightest bit of irony that they "could create a joyful, organic atmosphere rather than the simple coming and goings of inorganic vehicles."[28]

More recent research aiming to read drivers' emotions to alter their mood within the capsule-like environment of the car is far more invasive. From Jaguar to Toyota, car manufacturers of all ranks are deploying controversial facial-recognition technology coupled with AI-based detection to read the faces of drivers and adjust elements such as climate control and ventilation, lighting, media displays, and even scent in an effort to remake cars into "tranquil sanctuaries."[29] Utilizing what Jaguar, a British manufacturer of sedans and luxury off-road vehicles, calls an "onboard AI psychologist," the company's mood-detection system aims to learn the preferences of drivers and their passengers, making recommendations to correspond with their moods.

Because it is relatively easy to deploy, facial recognition is poised to explode in auto use, along with its well-known encoded forms of bias against non-white skin colors.[30] The names of numerous companies like Affectiva, Guardian Optical Technologies, Eyeris, Smart Eye, Seeing Machines, B-Secur, EyeSight, Nuance Automotive, BeyondVerbal, and Sensay all demonstrate how the lines between human and machine sensing and control are increasingly blurred, as well as the complex role these technologies have in sensing and tracking, scanning, and monitoring drivers and passengers as they sit, almost imprisoned within the car.[31]

It is no secret that some of the FAANG companies, as well as Microsoft, are working hard to exert their strength in the car sensing domain. Apple, for example, has filed for patents to use facial recognition to identify drivers in order to unlock cars, while other high-end manufacturers like BMW are exploring it to enable drivers to "navigate vehicles with their eyes."

Driving with your eyes is evidently not enough for BMW. The Munich-based auto manufacturer also imagines facial recognition (again) reviving the age-old concept of natural interaction that developed in the heyday of ubiquitous computing and embodied interfaces in the 1980s, creating a new technological approach within the driving arena.

BMW's own take on natural interaction not only includes paying attention to the road and other cars in terms of passenger safety and ergonomic comfort. It also involves autonomously detecting facial expressions to analyze which billboards, restaurants, or stores your eyes might land on as you drive and to conveniently order, purchase, deliver, and ship things you might briefly imagine having while driving by.[32]

These continued efforts to "humanize technology" have caused researchers and manufacturers interested in mood regulations to explore every possible angle. Real-time physiological signals from worn sensors, such as heart rate or sweating, are being harnessed to measure arousal and, increasingly, being paired up with neural networks to predict the emotional state of a driver. The results of these techniques range from altering the structure of music playlists to lighting up the car's interior with thousands of LEDs to visually display a driver's emotional mood to other drivers. With so-called Real-time Emotion Adaptive Driving (R.E.A.D.), Korean car producer Kia goes even further. R.E.A.D. aims to "amplify your joy" by reading passenger biosignals and combining these with machine learning to affect vehicle interior air quality, temperature, lighting, and even the opacity of the windows.

Light, color, and sound are the predominant media used in such sensor-driven mood enhancement, but scent is also on the horizon. An Israeli start-up called Moodify, which manufacturers "empathic car systems," is working on the delivery of "functional fragrances for improved safety and well-being."[33] Instead of pacifying the driver's mood with an ambient mix of new-age sounds, blinking lights, and vibrations like other research projects envision, Moodify's use of face recognition and AI aims to do the opposite: suddenly awaken drowsy drivers with noxious scents.

Moodify's founder, an Israeli smell researcher named Yigal Sharon, claims that smell—particularly, unpleasant smell—goes straight to the brain's equivalent of the jugular vein: the limbic system or primitive part of the brain that reacts to such stimuli as smell within a fraction of a second. The company's "vehicle olfaction" system is still in its infancy, but it perfectly bridges pleasure and security, the two visions of automobile sensory enhancement. That such sensor-AI-coupled technologies will most likely end up first in high-end autos suggests that your Chevrolet or even Smart car won't be seeing these mood-enhancing systems anytime soon. But the fact that much of the motivation for mood-changing research in automobiles is also being driven by the risk-averse insurance industry in order to avoid payouts on accidents caused by fatigued drivers also intimates that sensor-augmented emotional capture in the name of safety and security may indeed be implemented in cheap cars faster than we imagine.

Mood enhancement is not just about changing your emotions to positive ones while you drive. Its ultimate aim is to manage and reduce the stress of driving itself. If falling asleep at the wheel is one of the major causes of automobile accidents, then aggressive driving, better known as *road rage*, also inflicts untold harm. A 2004 US Department of Transportation National Highway Traffic Safety Administration study defined aggressive driving as "the operation of a motor vehicle in a manner that endangers or is likely to endanger persons or property" and declared road rage a criminal offense.[34]

The stress that leads to road rage during driving manifests itself in multiple ways: a cell phone ringing, traffic delays, shifting weather, noise, the distraction of accompanying passengers, and the aggression of other drivers. It is therefore hard to imagine how the sensors in a car could reduce all of these myriad influences in order to combat a phenomenon that in the United States alone experienced an almost 500 percent increase over the ten-year period between 2006 and 2015. If the latest research is any indication, the car as sensing machine will soon be charged with a new and needed task: providing stress relief during the act of driving.

Sensor-based stress detection lies somewhere between magic and science. The very concept of stress is difficult to define, given that it is something of a subjective state. The classic definition comes from the medical doctor Hans Selye, who described stress as any "non-specific response of the body to any demand for change."[35] Because of the ambiguity of *nonspecific* and the fact that stress appears in both physiological and emotional ways, sensing it is thus something of a challenge.

Stress is normally detected by specific physiological measurements such as heart rate variability (HRV), muscle tension, breathing rate, and arousal in the sympathetic nervous system that produces sweating. Each of these physiological signs is called a biosignal because it has a biological origin, and each signal has a specific sensor technology that can pick it up. Galvanic skin response (GSR) sensors measure the electrical or electrodermal activity (EDA) of the skin, which changes due to sweat produced by arousal triggered from the autonomic nervous system. Similarly, arousal can also be detected by electromyogram (EMG) sensors, which measure the changing electrical activity that occurs when you move or tense your muscles.

Now consider the fact that the physiological changes these sensors pick up only tell half of the story. Their readings then have to be contextualized to potential conditions. GSR sensors, for example, indicate changes in skin temperature that affect the electrical activity of the skin. It is much more difficult, however, to infer whether such sweating actually indicates the presence of stress, excitement, or some other factor. Just as Ernst Dickmanns's autonomous car needed many different sensors to perceive its environment, stress detection also relies on multiple sensing modalities to obtain an accurate physiological portrait of the on edge driver at the wheel. This factor, needing to know something about the context of where the sensing happens, is critical for accurately sensing any phenomenon. In other words, the context of the situation determines what can be measured and why.

But physiological sensors perceiving changing blood flow, breathing rate, and skin conductance have one major disadvantage: they all have to be close to the skin to measure these physiological signs and, hence, need to be worn. The context of driving makes such sensing bodily conditions rather unrealistic, being difficult, time-consuming, and expensive. Researchers are thus set on developing technologies to detect stress in less invasive ways through so-called sensorless sensing and to work at using this data to alter the surroundings that supposedly cause stress inside the car.

One technique is capturing stress markers from what are called *noninvasive technologies*, things like cameras, car seats, or other infrastructural elements like the steering wheel or gas pedal that one is in constant interaction with and that do not have to make direct contact with a driver's body. Cameras, for instance, are one obvious source of data that can be installed as "unobtrusive" sensors and can gather a flood of information from drivers, ranging from pupil dilation and eye blinking to more unexpected signals like facial temperature and percentage of eye closure.

More experimental techniques are also on the horizon. Computer scientists at Stanford University are "repurposing existing infrastructure embedded in modern cars" to measure the changing muscle tension that occurs as one grips the steering wheel at different angles.[36] Inducing varying levels of stress in laboratory test subjects by blaring heavy metal music from a Spotify playlist, the researchers modeled their test subjects essentially as machines, calculating arm muscle tension as a set of springs and examining how this tension changed based on the strength with and angle at which they gripped the wheel. Other techniques seem to come from gym treadmills: installing sensing electrodes in car seats, the steering wheel, and even seatbelts to gather electrocardiogram (ECG) data about a driver's pulse and heart rate in order to classify stress.

These techniques still don't address a critical and unanswered question: What does this data collected from these sensors actually do, and how is it used to relieve stress? To suddenly view the vital signs of your own physiology displayed as colorful graphics on the car dashboard like many engineers, not to mention graphic designers, working for the automotive industry envision would most likely induce even more stress. Imagine being confronted with an oscillating graphic of your heart rate speeding up or your hyperventilating breath chart zigzagging around while you're engaged in maneuvering through dense traffic.

Researchers therefore are looking for more subtle, ambient uses of data that can reveal changing physiological conditions. Like artists who use the movement data of their audiences to alter media in the surrounding environment, perception engineers are also exploring how to subtly but effectively shift the media environment of the car while you drive—and, with that, the function of driving itself. They even have a name for it: *environment soothing*.[37]

What does this soothing actually entail? For one, the car can now become an automated therapy device, with an almost seamless path between your interior psychic world and the car's. Sensors detect stress levels, and the data is then combed through by machine learning algorithms that have been pretrained to classify the level of stress and match it to an appropriate output. Adjusting the media of the car's interior environment then quickly follows. The color of the light changes. The sound adjusts itself.

The climate control shifts temperature. These transformations are all in the name of strengthening the perception-action loop between you and the rolling sensing machine you are (by now, only partially) steering.

One group of German researchers is even working with BMW to reimagine the car as a new "emotional feedback companion" that deploys a "complete ubiquitous sensing environment" within the interior to create a full-blown sensing and feedback loop. The use of "psycho-physiological" sensing, including brainwaves and heart rate sensing for emotion detection, gathered from the standard approach to worn sensors, would only be an input to a larger feedback loop between the driver and the color and intensity of ambient light inside the car, which can be shifted between blue and orange where these colors bring on different feelings.[38]

Environmental soothing might also result in receiving a gentle massage from the seat as you drive—a kind of twenty-first-century version of Magic Fingers, the coin-operated massage bed that was ubiquitous in middle-class hotels and motels in North America in the 1960s and 1970s. In car seats, for example, detecting unhealthy levels of stress could release soothing vibrations and warming heat within the same location. The technique is currently being examined for use in another artificial interior that some 4.5 billion people spent their time in during 2019: seats in airplane cabins.[39]

It might be surprising to think that more ubiquitous sensing technology can resolve the strain and anxiety of the daily commute. It could be better to reduce climate-destroying driving entirely. But the imagination of engineers shows no signs of abating. Environmental soothing involves a complete rethink of what we do in our cars while driving. The same group of Stanford researchers responsible for the stress measuring steering wheel is working to define a new concept for the therapeutic car: just-in-time (JIT) stress reduction. JIT is exactly what the name implies: solutions that mitigate stress at "just the right moment."[40] Imagine that moment. You are stuck in traffic. The symphony of horns from other irritated drivers begins to blare. Your blood pressure rises and your breathing speeds up.

Now consider the Stanford team's approach, which emphasizes "mindful stress reduction" and puts forward "subtle, unobtrusive, and easy to engage and disengage" interventions. In other words, driving can become a meditation session as well. Your irregular breathing, rise in blood pressure, and increased sweating are picked up by physiological sensors. After they detect these changes in your vital signs, the seat begins to vibrate. But it doesn't just buzz like your mobile phone. Sophisticated motors that are distributed throughout the seat start to massage you, attempting to influence the rate and expansion of your breath. The haptic guidance system is complemented by voice, which issues commands in the droll synthetic voice of one of the dozen voice samples in an Apple computer. The male voice instructs you as you begin to breathe in the same rhythm as the vibrations in the seat: "Breathe in. Now, breathe out."

Reenvisioning the car as a meditation environment perhaps goes too far. For now, these visions and technologies are still held back in the research laboratory, but they will probably soon find their way into most Fords, Nissans, Hondas, and BMWs. These research imaginaries from electrical engineers, computer scientists, and designers should not be ignored. They thoroughly demonstrate that even the daily grind of driving can be engineered into a new kind of aesthetic experience—a form of extreme interaction between the driver and the car's environment, which might result in a slight alteration of consciousness and maybe even another sense of self as we realize we are not separate from the (artificial) environment. These visions show that remaking the car as a sensing machine is not only about driving; it is about creating another outlet for engineered experience that involves continuous and unabated monitoring of our bodies and selves.

The future vehicle will leave no sensor untouched and no body unaffected. In this way, the car as a new domain for computer-augmented perception and action has dramatically evolved since Ernst Dickmanns's vision of an autonomous vehicle navigating the world with dynamic vision. It is no longer simply a driving machine to get us from point A to point B. With the ability to perceive and act on our surroundings and us, the car has become sensing machine *à la lettre*: a new set of interchangeable, almost immersive environments that can switch between spa, office, club, zendo, clinic and most possibly, prison.

7 Tasting Machines

Tastes are made, not born.
—Mark Twain

In the 1990s, a little-known Japanese engineer was working as a research assistant when he ate a life-changing meal. He didn't particularly like carrots, but his wife, being concerned about his health due to his overworking himself, chopped up the vegetables into tiny, almost indistinguishable pieces and folded them into hamburger meat. The engineer suddenly wondered why an ordinary hamburger steak could taste "so delicious" and was overwhelmed by "being allowed to experience the mysteries of taste."[1]

Kiyoshi Toko's research revealed an unusual fact. While there were sensors for vision (light and cameras), sound (microphones), touch (pressure and temperature), and smell (gas), no one at the time had developed technologies to artificially taste. Toko, who now runs the Research and Development Center for Five-Sense Devices at Kyushu University in the south of Japan, set out to defy conventional wisdom that saw taste as such a fundamentally subjective experience that no artificial device could ever hope to replicate it. But he reframed the issue, from one of taste being unique to an individual's genetic and cultural makeup to a neuroengineering problem: taste as a specific set of nerve reactions.

Toko set about developing technology to mimic a biological trait—an array of chemical sensors that could detect bitterness, sourness, or sweetness when the device's sensors came into contact with different chemical molecules. Toko's "e-tongue" simulated a biological material, a *lipid*—an oily and waxy substance that makes up the building blocks of living cells—in order to simulate the taste buds' reception of food converted into fluid by the relentless brutality of the tongue, palate, teeth, and jaw.

Toko quickly realized, however, that sensifying taste was not a trivial thing. Taste consists of synthesizing hundreds of different molecules through the taste receptors, but in the end we are left with only five choices. Our human tongues don't break down complex tastes, like coffee or chocolate, so that the brain can pick out individual molecules. Instead, we just pronounce a food as tasting salty or bitter or any of the other five tastes.

Researchers call this phenomenon *selectivity*—the ability to decompose hundreds or thousands of molecules into individually recognizable substances. Each taste bud has around 50-150 receptor cells that food molecules bind to and that represent the five tastes. In other words, even though the amount of chemicals that enter the mouth when we eat is astoundingly large, our mouth and brain cooperate with each other to deduce the five basic tastes out of a molecular chaos.

Toko and his research team concluded it would be impossible to have a different sensor for every taste molecule. Instead, he addressed the selectivity issue by developing hardware and software that would exhibit behaviors similar to human physiology: the ability to break down chemical substances into the five tastes, or what he called *global selectivity*.[2]

The Japanese researchers argued an important point in conceiving how to translate human taste sensing into a machine. *High selectivity*, the distinguishing of individual molecules, in the mechanism of the sensor is not important. The software that holds pattern-recognition systems that classify thousands of molecules into five basic tastes is where taste is really located.[3]

What is more extraordinary is that Toko's e-tongue was developed before scientists discovered the existence of the individual receptors in the taste buds in the early 2000s. The e-tongue technology was commercialized in the mid-1990s and quickly proliferated internationally in the factories of the industrial food titans, mainly used in quality-control applications. But e-tongues have also exhibited more remarkable abilities than their humdrum duties detecting food spoilage. They have been trained to discriminate between different types of coffee and a wide ranges of beers. These sensors can also detect more exotic, highly subjective sounding things, such as "richness" or even astringency.

Tasting Machines

The very notion of an electronic tongue brings to mind the image of a grotesque prosthetic organism straight from a David Cronenberg film. Disappointingly, industrial e-tongues look nothing of the sort. Instead of a fleshy mutant-like form, they appear more like a cross between a small robot and a food processor, equipped with a sensor-mounted arm that is lowered down into small cups on a rotating base that feature different samples of liquid to be "tasted."

Despite their less than human appearance, e-tongues are nevertheless starting to replace human sensory tasters, especially in some more unusual applications. Toko's research team discovered in 2014 how to detect the differences between artificial sweeteners, together with a slightly more bizarre revelation: e-tongues can taste flavors that are not present. In one set of experiments, the e-tongue discovered a hidden "third taste" based on the combination of two base tastes of mainly Japanese ingredients. For example, an e-tongue sampling oolong tea and carbonated water together would

Figure 7.1
Electronic tongue in use at the US Department of Agriculture. Photo courtesy Stephen Ausmus/USDA.

indicate that it tasted beer, while yogurt and tofu suggested the taste of a Japanese-style cheesecake.

Whiskey seems to also be a favorite drink for taste sensors. Toko's research led to the development of a Japanese highball cocktail combining whiskey and soda, which was subsequently served on the ANA airline. In 2019, a group of University of Glasgow researchers created a new generation of artificial tongue: specific nanoscale sensors that could detect the differences between different ages of scotch at 99 percent accuracy. While the scientists suggest that such a nanoscale technology could be used in the standard applications of quality control and security, they also proposed that the sensors could be useful to control the market in counterfeit whiskey.

In mid-February 2020 before international restaurants succumbed to COVID-19 shutdowns, a physicist friend and I visit Tickets—the first Barcelona restaurant opened by Albert Adria, the brother of Ferran Adria, perhaps the most famous chef in the world and former creative director of the now-closed experimental elBulli restaurant north of Barcelona.

Tickets is about as far from the clinical environment of Toko's Kyushu laboratory as you can imagine. It is, for all intents and purposes, a tapas bar specializing in something called *modernist cuisine*. Over the course of three over-the-top hours, a parade of small dishes continually appears from the kitchen: spherical olives; the "air baguette" (a crunchy hollow baguette wrapped in a slice of Rubia Gallega beef); a delicate potato puff with mustard seeds; a roasted quail stuffed with ham and herbs, wheeled out on a cart and carved table side (see figure 7.2).[4]

Figure 7.2
Tickets's sensory taste experiments. Photos by author.

Most of the dishes comprise only a moment's gustatory pleasure, and then they are gone. Small bottles of Estrella beer continually appear, along with the occasional pair of tweezers to sample paper-thin shaved squid and minute morsels of sea urchin. During the course of the hours-long meal, we watch Albert Adria in the open kitchen. Tonight, he doesn't cook himself but instead gazes intently into a hanging LCD monitor—presumably overseeing the complex choreography of courses being ordered, cooked, finished, and served. After several hours, the waitress somehow asks almost innocuously if we are ready for dessert.

Dessert takes place in another room that we are guided to. The cataract of molecular transformations is never ending. The lemongrass air popsicle. A narrow ice cream cone that is frozen by liquid nitrogen. The iced foams and hazelnut iced cheesecake. Dessert seems to last as long as what came before. On the back wall of this fantastical chamber with reams of hanging plastic fruit and gold lamé–painted ceiling is a rack of dessert cookbooks from the world's most illustrious pastry chefs. In a word, this is gastronomic overload.

It would seem impossible for e-tongues to "taste" the complex parade of gustatory pleasures emanating out of Adria's alchemy-like kitchen. Indeed, of all of the senses, taste seems the least likely to be effortlessly replaced by machines. It is deeply evolutionary, tied to the most primitive biological urge to feed oneself in order to survive—an urge neither electronic sensors nor machine learning possess. "No matter how cultured one's palate or subtle the ingredients in a dish," writes Pulitzer Prize–winning journalist John McQuaid in his study *Tasty: The Art and Science of What We Eat*, "a taste summons raw urges out of the deep past, echoing evolutionary twists and long-ago life-and-death struggles over food."[5]

Taste is deeply sociobiological. It is what researchers call *multimodal*, interfered with or changed by other senses like smell, touch, and hearing. Taste is also culturally and socially learned. While our aversion to the flavor of cilantro can be attributed to genetic structure and the primitive amygdala part of the brain, what we choose to put in our mouths is equally shaped by where and how we are brought up.

But this sense by which the flavor of things is perceived as they are brought into contact with the tongue poses an even more fundamental challenge to sensing machines. Taste is fiendishly difficult to standardize and render into machine-readable numbers. Even though animal and

Homo sapiens mouths break down the structure of food into watery particles, thus exposing the taste buds to thousands of molecules and allowing them to distinguish between broccoli and dark chocolate, none of us perceive these flavors exactly the same.

Taste, in other words, is a sense that works in mysterious ways. Long thought to be controlled exclusively by the taste buds, its story has more recently been updated to reveal richer complexities. The between two thousand and ten thousand sensory receptors on the tongue and in the mouth that we call taste buds are equipped with ten to fifty taste receptor cells. Within each of these receptors are proteins that discriminate between the five tastes: sweet, sour, bitter, salty, and the most recently named fifth one, umami (savory). Some taste buds are more sensitive to sweet things, while others excel in identifying a salty food, but every taste bud is equipped with receptor cells to identify the five core tastes.

These cells activate groups of nerve cells that chemically differentiate combinations of molecules. Finally, signals fired from these nerve cells are sent to the brain, which then categorizes them into one or sometimes more of the five categories. Never tiring from this ceaseless analysis performed on everything we eat, the taste buds have a trick up their sleeves: they frequently die, only to be freshly generated every one to two weeks.

Molecules exciting the taste buds are not the only entities responsible for why something tastes a certain way. Temperature, texture, density, and, most of all, smell, another set of molecules, also radically influence our gustation. According to Gordon Shepherd, a neuroscientist studying the neurological basis of flavor who proposed a new field of research called *neurogastronomy*, smell makes up (at least for the brain) most of what we call *taste*. Sniffing food doesn't consist of breathing in but of breathing out smell onto food, a process called *retronasal smelling*, or simply *mouth-smell*.

There is also another entity that lurks in the mouth and that also accounts for taste. A facial nerve running from the lower jaw into the mouth also permits us to taste things that don't operate in the classic five tastes sensation hierarchy. The trigeminal nerve is responsible for our sense of chewing, but certain foods trigger a response as well. Trigeminal sensations are thus the reason that some people feel cilantro or coriander tastes soapy, or why chocolate leaves a rough or astringent taste sensation in the mouth.

That taste intertwines evolution and culture, nerve cells and molecules, physiological chemistry and genetic code, makes it a particularly important research field that has long been ignored not only by medical researchers but also by social scientists. But there is another reason that taste is quickly becoming a new category for sensor-based intervention: it is worth a lot of money.

The global processed food industry lives and dies by this fact. More than any other industry, industrial food production has turned the engineering of artificial taste and smell into a $2.7 trillion annual revenue stream. Artificial flavors like neotame, a sweetening agent that is eight thousand times sweeter than sugar, or diacetyl, an infamous carcinogenic butter-tasting substitute formerly ubiquitous in microwave popcorn, confound the senses and bludgeon the brain into thinking they are something they are not. The substances that make up "natural flavors" derived from plant or animal sources such as fruit, meat, fish, spices, herbs, roots, leaves, buds, or bark and that are distilled or fermented in the lab are no match for the estimated 90 percent of artificial taste substances embedded in processed food on American supermarket shelves.[6]

Yet flavor, either natural or artificial, is a slippery concept. Flavor is not really contained within things, neither food nor molecules nor the sensory cells of taste buds, but, as neuroscientists argue, lies between them: in the neurochemical information trail that bridges molecules, the nose, the mouth, and the brain. At the same time, sensory anthropologists whose expertise now expands to study the human senses as sociocultural phenomena point out that flavor is not just born, the result of genes and neurons, but is also made. It is the product of culture itself.

None of these scientific disputes alter sheer economic reality: flavor is a chief reason that consumers purchase one food brand over another. Industrial food giants are at perpetual war for our mouths and dollars in the battlefield of the local supermarket, and flavor is their ammunition. It comes as no surprise then that there is relentless pressure for perpetual innovation in the universe of artificial food texture and flavor. The brutal market for processed food producers to maintain competitive advantage has resulted in a technological tsunami of invasive interventions into what we eat and devastating health-related consequences worldwide as a result of these engineered concoctions.[7]

In 2017 alone, the market for *taste modulators* and products exploiting *mouthfeel*, two core, state-of-the-art research areas in flavor engineering, was estimated to be worth almost $1 billion USD and growing annually.[8] Modulators are chemicals added to food that improve or mask (hide) the quality of its taste. "Healthier" salt and vinegar potato chips that are made with less sodium can miraculously taste saltier. The bitter taste of a sugar substitute like stevia can be masked to appear sweeter than it is. By precisely adjusting food's molecular composition, chemistry reverts back to its alchemical origins. Artificial ingredients are the new base metals awaiting transformation into gold.

The research area called mouthfeel is even more revolutionary. Mouthfeel is what produces the sensation of why one kind of potato chip seems crispier over another or allows us to distinguish degrees of creaminess across different brands of instant pudding. Describing the texture and feel of food inside the mouth, mouthfeel permits us to announce that a food tastes sticky, gummy, chunky, mushy, crispy, velvety, or chalky.

If the taste buds and nerves are focused on chemical sensations, mouthfeel is something else: a purely mechanical and acoustic sensation related to how our own fleshy pressure sensors, the tongue, the palate, the teeth, and the jaws, chew and process what enters our mouths. Mouthfeel is the physical sensation of food.[9]

The need to tune and tweak every last molecule in cake mix or cream cheese has, not unexpectedly, spurred the rapid evolution of a parallel research area called *sensory science*, or its earlier incarnation, *sensory evaluation*, dedicated to quantitatively measuring the sensation of taste. As a standard textbook definition puts it, sensory science "comprises a set of techniques for accurate measurement of human responses to foods and minimizes the potentially biasing effects of brand identity and other information influences on consumer perception."[10]

This artificial engineering of food is fundamentally intertwined with the human senses of taste and smell. To eliminate human bias from its experiments, sensory science is conducted in the clinical confines of the laboratory. In the cloaked secrecy of corporate flavor conglomerates in Switzerland and suburban New Jersey, or independent organizations such as the mysterious Monell Chemical Senses Center in Philadelphia, armies of psychologists, physiologists, biochemists, and psychophysicists study how smell and taste can be calculated.[11]

These experiments are by necessity reductive. They eliminate the potential influences that could bias taste: the color of the label, the name of a brand, the claims and promises on a box. Their experimental arrangements are designed to thwart the senses. Taste testers are put inside closed cubicles, socially distant from each other. They are blindfolded or have their ears stuffed with cotton to eliminate the possibility of sound and sight interfering with experiments. Taste samples are provided in generic containers marked solely with numbers: 1, 2, 3, 4.

The pleasure of taste is converted into numerical categories in order to "establish lawful and specific relationships between product characteristics and human perception."[12] These relationships are analyzed and conclusions drawn from them that should result in potentially new products or modulations of existing ones.

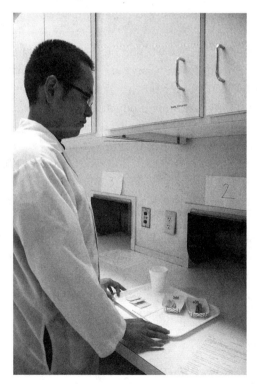

Figure 7.3
Blind taste test in sensory science laboratory. Photo by Katie Walker.

Sensory science is the ideal mix of hedonism and measurement. It is none other than psychophysics, the discipline that Gustav Fechner invented in the mid-nineteenth century that quantitatively measures the perceived intensity of a stimulus. Apply psychophysics to the evaluation of taste and smell and you have it: a cornerstone in the production of industrially processed food. Everything that provides us pleasure in our encounters with food can thus be subject to measurement: the bitterness of chocolate, the texture of a new cookie, the thickness of an emulsion for mayonnaise, or the way that fat tastes as it melts in the mouth.

But this science of measuring taste is not just about difference; it's also a guarantee that batches of food coming off the conveyer belt taste exactly like other batches. Ensuring the standardization of taste is one of the reasons that the world's largest industrial food production conglomerates—Nestlé, ADM, Cargill, and PepsiCo, among others—are on a continual hiring spree for sensory scientists. Consider this description on Cargill's website: "Cargill's comprehensive capability in sensory science and technology helps customers develop new products and maintain consistent quality and 'sameness' in their existing products. Cargill now offers a superior ability to predict taste sensations—a capability that helps identify the natural molecules in flavor ingredients that enhance taste and ultimately lead to more appealing consumer products."[13]

How then does sensory science do its science? Quite simply, by way of human sensors—panels of trained tasters and sniffers who use their finely tuned noses and taste buds to discriminate between different samples; describe varying perceptible characteristics like color, texture, or flavor magnitude (intensity); and then quantifying their degree of like or dislike, what is known as rating the affective quality of the product.

These human sensors not only tell the Campbell Soup Company that its new chicken broth needs to be saltier. They also become a new kind of invisible, unacknowledged labor, assert Christy Spackman and Jacob Lahne, two social scientists studying the sensory impact of food—a kind of labor in which the body doing such sensory work is essentially "erased."[14]

Sensory science sees the human body as its measurement instrument. One recent textbook is blunter. Human testers are "heterogeneous instruments for the generation of data."[15] Canadian anthropologist David Howes, an

Tasting Machines

expert on the sociocultural context of the senses, calls this a paradox. The core mantra in the sensory science industry is that no machine can replace the human senses. Yet "it is considered important that human subjects behave as much like scientific instruments as possible."[16]

But the days of humans blind taste-testing industrially produced Oreos and spaghetti sauces are numbered. E-tongues and other emerging sensing machines are rapidly demoting human taste testers. Large-scale automated sensing systems combined with the computational firepower of statistics and machine learning are also remaking industrial food production and the tasting that goes with it. Evaluation processes are being automated at scale, making the erased labor of the human tasters even more invisible.

Sensors in industrial food production are ubiquitous. They are attached to conveyer belts, mounted on robotic arms and flying drone inspectors, used in batch processing, and manually handled by assembly-line workers armed with handheld devices analyzing temperature, flow, chemical composition, mass, and volume. Sensing has advanced knowledge in an obscure branch of physics called *rheology* that studies the deformation and flow of materials, helping to better understand why ketchup gets stuck in a bottle or how batter can cling better to fish sticks. Researchers have even used computer vision and, increasingly, machine learning to solve food technological questions that to most of the world's population must seem downright irrelevant. In 2007, for example, researchers deployed the computer vision technique of edge detection to examine why the texture of pizza dough degrades over time.[17]

The use of sensing machines in industrial food historically focused on maximizing quality control and process efficiency. The redness of tomato pastes or the shine of hard candy has been measured by spectrophotometers and refractometers. These devices sense color reflectance and light scattering in order to detect whether the material properties of substances are correctly structured. Infrared thermometers and thermal-imaging cameras process images using the mathematics of black bodies in astronomy in order to monitor the degree of thermal radiation given off from warm loves of white bread. Ultrasound emitters can calculate the speed of bursting gas bubbles in Coke and beer.[18]

The majority of these sensors are based on earlier electromechanical or acoustic technologies. But these devices' days are numbered as well, superseded by the next generation of so-called biosensors. Biosensors' sensing elements, their receptors, are closer to real biological sensors—tongues and noses, or the touch sensors embedded in our skin. These organic sensing elements, like materials made of biological substances such as enzymes or even antibodies, are then connected to electrical components called *transducers*, devices that convert one kind of energy into electrical voltages that a computer can read.[19] From antigens and antibodies to chemical reaction and microorganism detection, biosensors are increasingly proliferating in the machineries of processed food making, much like the very bacteria they aim to detect.

However advanced it is, the processed food industry came relatively late to the sensing game. According to a 1997 *New York Times* article, the automation industry in the 1980s sought out new markets for machines that had previously been employed in cyclical or downsized industries, like petroleum monitoring or shipbuilding. To an industrial viscosity sensor that at one time measured the thickness of refined oil, a new application like monitoring the creaminess of salad dressing would be no big deal.[20]

But it's only since the 1990s that industrial food production has sought to use sensing systems at scale to replace what human tongues and noses formerly did. As we see from their commercial deployment, e-tongue sensors are now sniffing and tasting away in the factories of Nestlé or Kraft Heinz, encouraging what one Kraft engineer called "a movement in the industry away from the art and toward the science of making food."[21]

E-noses and tongues are a critical component of this new food science. Yet discriminating between the intricate amalgams of chemicals that make up concentrated orange juice or sprayable cheese in a can is not an easy task for machines. This complexity is one core reason that the development of sensors to detect odors and tastes is a more recent phenomenon. Unlike hearing, seeing, and touching, which respond to mechanical and electromagnetic stimuli, smell and taste are chemical senses that rely on the detection of molecules in the air or in the mouth.

Take, for example, the e-nose. E-noses were developed in the food industry in the early 1990s to cost-effectively monitor taste in test situations. These olfactory doppelgangers consist of a tangle of sensing elements and algorithms—clusters or arrays of small chemical sensors together with microprocessors and computational models that sort and classify what the

Tasting Machines

sensors are "smelling."[22] E-noses are groups of interlinked gas sensors that detect molecules. Like the human nose, they analyze aromas by detecting, or what scientists call *registering*, signals produced by a mixture of gases, and then they compare the pattern of responses.[23]

But how do these artificial noses actually smell? Because detecting molecules is a chemical process, artificial noses have special electrical components called *chemoresistors*. These components influence the flow of electricity within a circuit based on chemical reactions happening on or around the sensor, similar to a standard electronic resistor.[24] Smell molecules interact with these components, causing electrical changes that are subsequently registered by software. Molecules thus become data for a machine.

E-noses need something else to tell what they smell: software. Statistical processes that learn to identify patterns in existing data are let loose on molecules to extract the specific characteristics of molecules, such as frequency (how fast they vibrate), phase (when the vibration starts and stops), and amplitude (the intensity of the vibration), that can tell mathematical models something about what the sensors have smelled. The mathematical models use sophisticated statistical processes with names like *classification* and *segmentation, regression,* and *dimensionality reduction* to do things that biological smelling systems innately do: separate groups of molecules to distinguish, classify, and name one smell over another. But mathematics has sensory limits. Not even these computational workhorses are remotely close to matching the human nose's ability to detect and classify upward of ten thousand different molecules with approximately only three hundred plus receptors in the olfactory bulb.[25]

Human noses are not the only olfactory organs that engineers are modeling. There is also extensive research examining and incorporating the sensing organs of insects and arthropods to detect the odor of rotting meat and spoiled milk. Arthropods, like insects, spiders, crabs, shrimp, and millipedes/centipedes, have highly sensitive chemoreceptive organs, encouraging researchers to reengineer and interface them directly into microelectrical arrays and processors.[26] The long-imagined future fusing the biological and the electrical in a circuit is already upon us.

Things are poised to get ever stranger. The advent of high-speed fifth-generation (5G) internet networks will launch a new era of sniffing and lip-smacking sensing machines. The prefix *tele*—the Greek word for distance—long preceded words that marked technological revolutions in sensory perception: writing (telegraph), hearing (telephone), seeing (television). Historically, writing, speech, sound, and sight at a distance were the dominant sensory media that compressed space and cut up time. According to molecular gastronomy physicist Hervé This, soon smell and taste will join as telesenses as well. "What realms of communication are left to conquer?" This asks. "The transmission of tactile stimuli is now being mastered, thanks to the design of special gloves fitted with piezoelectric crystals that register or exert pressure. But smells? Flavors? The delay in developing tele-olfaction and telegustation is indeed a source of frustration for gourmets."[27]

Where does all of this artificial sensing of taste and smell lead us? Will we taste and smell differently by sharing these duties with other devices? Are we developing machine tasting counterparts that tune our food before we actually taste it? E-tongues may have already decided the exact amount of tuned bitterness a new microbrewed beer you purchased has before it ever hits your mouth. When your favorite soft drink changes its secret formula, your tongue may first pronounce harsh judgement, but the electronic sensor arrays may already be hard at work to test and taste the back to the drawing board results revealed by your human evaluation. Is it replacement of the senses, or cooperation—tasting *with* machines?

For electronic tasters and smellers, gustation and olfaction are reframed as statistical detection, analysis, and classification. The pleasure of eating and the satiated feeling of sustenance in the stomach drop out. Mathematics now sets the parameters—parts per million, speed of sound, radiation level, torque, viscosity, shear stress, and substrate mass. The quantification of chemical substances and their segregation to generate machine-readable patterns is the reinvention of taste and smell for sensing machines. Gastronomy, the study of the "laws that govern the stomach," has reached its scientific quintessence[28] It is recast as numbers and equations.

But not so fast. E-tongues might find applications where such calculation that transcends human capabilities is actually useful. It seems problematic that the next generation of e-organs would only be in the hands of Nestlé or Coca-Cola. This is about to change. Imagine a near-future scenario in which we use these emerging sensing machines as a consumer weapon against the health-damaging processed food colossals—in which we consumers have access to e-tongues in our own kitchens that can taste food and sample drink we have bought.

These sensing machines could quickly identify lurking chemical dangers in foods that, at least in North America, are innocuously labeled as containing "artificial flavors" without specific details of their chemical substances being given. They could be useful in identifying foodborne illnesses, an area that, according to the WHO, results in some 420,000 deaths annually.[29] E-tongues could detect poisonous substances in spoiled or rotten food before we sample it. This puts these prosthetic organs at a distinct advantage over our fragile mortal coils.

There is already a hint of such possibilities soon becoming available to us. Using a different sensing technology formerly in the hands of research scientists, consumers will soon be empowered to do their own nutritional detective work in the produce section of the local supermarket. A smartphone-sized sensing device that uses a technique called *near-infrared microspectrometry* enables the device's sensor to absorb light reflected from objects and can thus instantly decompose this information to understand a fruit or vegetable's chemical makeup, with the data all delivered straight to a palm-sized gadget.

In fact, if these artificial sensors already transcend some human tasting or sniffing abilities, why even use biological terms like *tongue* or *nose*? After all, e-tongues and e-noses have no intrinsic understanding of why they taste or smell what they do—concepts that are intimately intertwined with human sensing. The triggering of memories based on smell or taste, so vital to human experience, are absent from these machines, as well as the social contexts that shape why, how, when, and where we eat and smell.

More to the point, taste and smell are intertwined to such a degree that the loss of one sense through accident or genetic error can result in a disappearance, mutation, or transformation of the other. As one UK whiskey expert, who dismissed the ability of the e-tongue to detect differences in scotch, argued:

> Flavor assessment in the whisky industry is done by smell, taste and texture. Of all the senses employed, smell is the most important. Whisky blenders and quality assessors rely entirely on smell . . . Our sense of taste is crude in comparison. While we have . . . around 9,000 taste buds, we are equipped with between 50 and 100 million olfactory receptors and can detect aromas in minuscule amounts, commonly in parts per million, sometimes parts per billion and with some chemicals, parts per trillion. Will this [e] 'tongue' be as sensitive as this? I wonder.[30]

IV Monitoring

8 Ten Thousand Steps

> Unless something can be measured, it cannot be improved. So we are on a quest to collect as many personal tools that will assist us in quantifiable measurement of ourselves. We welcome tools that help us see and understand bodies and minds so that we can figure out what humans are here for.
>
> —Kevin Kelly, writer and cofounder of the Quantified Self movement

On February 9, 2018, there were a lot of disappointed Spotify users. The Swedish music service announced that one of its most popular services, called Running, was to be retired. Debuted in 2015, Running was a special feature that aimed to create a seamless connection between a runner and their music. Spotify's algorithm, fed by a smartphone's accelerometer and gyroscope, searched a track to identify and play only the "high-energy parts of each song" and then cross-faded to another, all without a break. "By the time you were sick of a song, Spotify was already moving to the next."[1]

Fans didn't take lightly to Spotify's decision to discontinue Running. Online social news aggregator Reddit was awash with complaints that the service's removal would be "a major loss for the platform." One Running fan explained, "The only criteria for whether to destroy a feature seems to be 'if it's useful and fun, gives a better user experience then it goes into the trash.'" Others incredulously asked, "Why would they develop such a cool feature and then just turn it off without giving any reason?"[2]

At first, this strategy of Spotify seems par for the course. Corporations are continually discontinuing technologies that consumers are addicted to. Spotify claimed that something better would be coming. The service would be "pouring our energy into new ways of creating the best experience for our users." It even offered a weak alternative: one could "still workout with Spotify standard playlists as your Running companion."

Reading between the lines, these Spotify user complaints perhaps revealed something else about their disappointment—the fact that many users were attracted to the direct coupling between their physical movement and the music's response. One user stated that the feature "was pretty cool the few times I used it . . . . The music would change and adapt as I progressed through my run. For example, I swear the music would crescendo/swell when it detected my pace was slowing down, encouraging me to pick it up."[3]

Perhaps Running was discontinued because it was merely a test case, an experiment Spotify unleashed as part of the company's larger ambition to position itself not simply as a music delivery service but a media lifestyle corporation that would create new affective links between music and a user's physiological and emotional states. Just glance at some titles and short descriptions for Spotify patent applications between 2015 and 2020: "Media Content Systems for Enhancing Rest," "Systems and Methods for Enhancing Responsiveness to Utterances having Detectable Emotion," sensing of user behavior with "a motion, hand/proximity, or heat sensor, which detects when a user is about to interact with the device" or, most likely the original research vision behind Running, titled "Physiological Control Based upon Media Content Selection." The patents all involve the similar use of detection devices and machine learning to load, buffer, select, and play audio. In the case of the physiological control patent, the aim is unambiguous: "to cause the media-output device to modify playback of the media content items based upon the current physiological measurement."[4]

When it debuted, Spotify's advertising for Running reinforced the emotional attachment users had in being one with their music. "You can soundtrack your entire life with Spotify. Whatever you're doing or feeling, we've got the music to make it better." In other words, Running was not just a Spotify feature. It was an attempt to forge a new kind of interactive experience between an exercising body and media environment (in this case, music) based on the expanded sensing and algorithmic capacity embedded in a standard smartphone. The phone became inseparable from one's moving, active body. The sensing machine was now reimagined as an arbiter of a new fitness-driven body and self.[5]

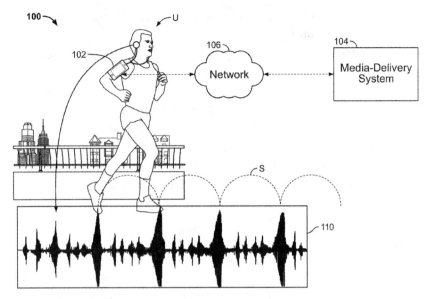

Figure 8.1
Example drawing for heart rate determination and media content selection system from Spotify patent.

Spotify didn't invent the concept of the exerciser "soundtracking" their life with a media device. That concept harkens back to 1979, when Sony released a small portable cassette tape player called the Walkman. Sony's smash commercial success wasn't just dependent on large numbers of teenagers who used the Walkman to listen to prerecorded cassette tapes while walking around with headphones oblivious to the outside world. The portable apparatus also gained relevance as the age of the aerobic workout emerged, capitalizing on exercisers' desires to have a personal musical companion while sweating it out in the living room.

Despite its myriad versions, the Walkman and its televisual successor, the portable Watchman, had no sensors to monitor your movement. The Walkman didn't know how hard you worked out, where you ran, or how many steps you took. In other words, the feedback loop between the exercising listener and their device's media playback was emotional but not driven directly by sensors and software. No matter how fast you ran or hard you breathed, the Walkman's sound would remain the same. The device was closed to the world.

Machines able to sense what is going on around them like the wearable fitness tracker and the smartphone, which are as commonplace today as the Walkman was in the 1980s, would only gradually appear. Part of this was due to the shrinking form factor of silicon and electronics, which allowed for the miniaturization of sensing. But these portable media devices were also awaiting a new cultural moment, one that envisioned a novel, embodied way of relating to computers. From hand gestures to full-fledged cardiovascular exertion, physical bodies in motion would become the new interface between mobile computing machines and equally portable media experiences.

Spotify's Running demonstrated this kind of thinking. Here was an application that directly sensed body motion with the phone's accelerometer and gyroscope and used this sensor data to choose, switch, and crossfade music all seamlessly, in real time. While Running could only function based on the number crunching enabled by the algorithms in the runner's phone, the user's direct experience was ultimately not altogether quantitative. It was qualitatively different than simply adjusting their workout behavior based on numbers appearing on a visual display, like *how many steps*, *today's* and *this week's activity*, or *goals met*. You could say that one's workout improved, even optimized, without the benefit of seeing any numbers. Instead, Spotify understood the power of sensing to create an emotional resonance between one's physical actions and the affective force of media—in this case, music. Running promoted a model of sensing machine in which a moving body, sensors, and the environment one was within could be imagined as a single interacting system.

Data scientists refer to the algorithms that drive systems like Spotify as *recommender systems*.[6] Recommender systems are mathematical models that try to predict user preferences based on a history of interactions with a particular system. One of the core scientific articles on these technologies described their value as able to "make good matches between the recommenders and those seeking recommendations."[7] In the case of Spotify and other services, like Netflix, the system takes user profiles and listening preferences, as well as the behavior of like users (a technology called *collaborative filtering*), as input factors that drive the recommendations.

Running certainly worked on this principal of algorithmic curation by using mathematics to sort, filter, select, and present choices to listen to. But Spotify's now defunct service also suggested a new process—one of algorithmically driven sensation in which the emotional resonance and beat of music could also be automated to match the peaks and valleys of physical sensations that joggers experience over time.[8]

⌒

The idea of putting sensors into a portable communication device to measure even minimal physical action comes from an unusual context. In the mid-2000s, a group of computer scientists at the Indian Institute of Science in Bangalore sought to "bring the poor into the information age." Aiming to develop a low-cost handheld device that would "bridge the digital divide," this Indian-manufactured machine, called the Amida Simputer (supposedly a name derived from *simple, inexpensive, multilingual computer*) had an ambitious vision. Acutely aware of both the rapid speed of IT development in India and the lack of even the most basic telecommunications infrastructure in rural communities, the Simputer brought telephone and internet capabilities together in one handheld object.[9]

The Simputer would enable rural Indians to access the internet and to conduct financial transactions, connect over email, and, even more radically, communicate through a custom-designed text-to-speech system called Dhvani—allowing illiterate users to use their local languages. Illiteracy, therefore, would "no longer be a barrier to handling a computer." The vision for the Simputer was a "low-cost device, to help every Indian citizen—especially those at the bottom of the pyramid—irrespective of gender, language, physical handicap, geographical location, caste or creed, literate or illiterate, to access the information superhighway."[10] In other words, the inventors sought a revolution in both access and use.

The Simputer attracted global media interest, including an article by science fiction writer Bruce Sterling that claimed the device was "computing as it would have looked if Gandhi had invented it, then used Steve Jobs for his ad campaign."[11] But the Simputer's vision was more ambitious than simply selling mobile computers. The device ran on the open-source platform Linux, which gave it almost the computing potential of a small PC at the time (the device also had handwriting recognition), as well as a

Figure 8.2
Left: The Simputer. Photo from www.simputer.org. *Right:* The core team behind the Simputer (from left to right: V. Vinay, Ramesh Hariharan, Swami Manohar, and Vijay Chandru). Photo courtesy V. Vinay.

smart card, and cost around $230 USD. While still prohibitively expensive in terms of Indian wage standards in poorer communities, the founders had imagined a model for how these groups could get access. Local community organizations could purchase the device and loan it out to members of the community for short periods of time, with each able to have their own personalized smart card to use the Simputer.[12]

More astoundingly, the Simputer became the world's first handheld communication device to feature an accelerometer, at least in its commercial release by 2004. The two-axis accelerometer (then called a "flip flop motion detector") would ensure ease of use, particularly for those whom reading menu-based interfaces posed a prohibitive barrier. Low-literacy or even illiterate communities could thus learn to navigate a communication device in a multimodal way through image, sound, and touch.

Just think about this for a moment. Almost six years before the ballyhooed appearance of accelerometers in Apple's iPhone, an Indian invention principally designed for rural farming communities already facilitated a new physical relationship with a portable communication device: flicking through pages, changing images from landscape to portrait, zooming in and out of images with simple taps on the screen, and even playing a simple onboard game called Golgoli.

The Simputer was ultimately a commercial failure due to low take up in Indian government and corporate circles. But the idea of a computer intuitively responding to its users by way of nonverbal interaction through sensors was far-reaching. The sensor was not simply a technical gimmick; it was implemented in the name of creating an intuitive, easy-to-use interface that would respond not only to textual commands but also to human motion.

Of course, changing an image from landscape to portrait is not the height of physical exertion. But the intuitive physical interface envisioned by a portable sensing machine like the Simputer set something else in motion. The idea of carrying around a device loaded with sensors to provide for nonverbal and non-text-based interaction with a small computer was both ahead of and in the general air of its time.

Around the same period as the Simputer's 2004 commercial debut, for instance, telecommunication corporations such as NTT in Japan and Nokia in Finland also began incorporating next-generation accelerometers into smartphones. These were the same MEMS-based accelerometers that had transformed car safety and that were about to fundamentally reimagine video games and their controllers.

Unlike earlier one- or two-axis devices, the new 3D accelerometers had three axes, which made them ideal for sensing motion in three dimensions (up/down, left/right, backward/forward), such as that produced by the everyday activity of walking. The 3D accelerometers would soon be incorporated into what is called an inertial measurement unit (IMU). The IMU was special in that it could calculate not only acceleration but also angular velocity, a key feature needed to measure not only acceleration but also position and orientation.

The new three-axis accelerometers also advanced the use of sensors in smartphones to create a further connection with the exercising body by implementing a device most probably dating back to Roman times: the so-called step counter or pedometer. The pedometer measures the distance between two points by calculating the steps one takes. Some sources attribute it to the Romans, who had a wheel-based object called a hodometer or surveyor's wheel that was designed to measure the *mille passum* or standard Roman mile when rotated. Others speculate that Leonardo da Vinci invented the pedometer, as evidenced by the inventor's notebooks, which reveal sketches for a lever-like contraption attached to a small wheel. Presumably, when a soldier moved, it would cause the lever to shift up and down and rotate the wheel. The device appeared designed to measure the distance walked by fifteenth-century soldiers.

Still others invoke the name of Thomas Jefferson, who was rumored to have brought a waist-driven pedometer back from Paris, where he used it to calculate his walking between historical monuments. Even English scientist Robert Hooke is claimed to have created the first pedometer (what he called a *pacing saddle*), mainly used by the British scientist and inventor not to mark distances but rather to sweat in order to purge his body of the forces that gave him headaches, vertigo, and nausea.[13]

Despite its military-scientific origins, the pedometer seems to have also captured a kind of popular status in the early to mid-twentieth century after being picked up by walking enthusiasts. In the US in the 1930s, long-distance walkers referred to it as a "Hike-O-Meter." In a run up to the 1964 Tokyo Olympics, Japanese walking clubs began enthusiastically purchasing a device called the *manpo-kei*, or ten-thousand-step meter.

The ten-thousand-step number is magical for fitness aficionados. It appears in almost every advertisement for fitness trackers, on wellness blogs, and in online fitness news groups and Facebook pages. In fact, the number is arbitrary. A young Japanese physical health sciences researcher named Yoshiro Hatano, now considered to be the first to link the pedometer to the idea of encouraging physical wellness through walking with a device, came up with this magical figure.

With a growing fear that in the postwar economic boom Japanese men were increasingly becoming sedentary and obese, Hatano authored a study to prove that a healthy lifestyle should include ten thousand steps (around five miles) daily, which would amount to around 20 percent burn-off of calories over twenty-four hours—a "golden number" quickly deployed as a catchy advertising slogan by the Japanese Yamasa Tokei Keiki corporation, which developed a kind of pedometer in 1965 to be marketed to an increasingly gadget-prone and gradually health-conscious Japanese public.[14]

The pedometer's widespread take up in Japan, where many senior citizens still own at least two of the devices, was not overlooked by industry. Thirty-eight years later, in 2003, the Nippon Telephone and Telegraph (NTT) Docomo division released its Raku-Raku ("easy to use") Fujitsu phone, which was the first to feature a pedometer. Aimed at seniors who were both familiar with pedometers and desired an easy-to-use device, the step counter in the handset was on 24/7, operating in a stand-alone mode unknown to the user. It could run continually, even when stuck in a pocket or deposited in a purse.

What's even more intriguing is that the Fujitsu phone had an integrated email application that could send messages to the user at specified time periods daily to indicate the amount of activity and calorie consumption recorded. With the Fujitsu phone, the pedometer made a radical transition, from that of a device needing to be turned on and off by its users to one involving a continuous monitoring of activity that remained invisible to and, increasingly, inseparable from its user. The pattern of activity itself and not only individual actions at discrete points in time could thus be used to both reveal something interesting about a user's physical actions and act as a catalyst to influence their physical behavior.[15]

With the goal of behavioral change in mind, a mere three years after the Fujitsu phone, Finnish firm Nokia released a competitor. The Nokia 5500 Sport phone was also equipped with a 3D accelerometer. While the step-counting feature quickly exhausted the device's battery life early on, the phone itself presented a future-oriented vision of sensors becoming instigators of yet another new consumer: the on-the-go, active lifestyle user. In 2006, more than thirteen years before Spotify Running, Nokia had anticipated that the phone would become an essential accompaniment to fitness. By shaking the device, for example, one could choose music while on the go, jogging, running, or performing a similar activity.

With its palm-sized form factor, image of durability due to its rounded corners, and rubber outlines and carefully cultivated image of the phone attached to the body as a necessary accessory (the device needed to be strapped to the waist for the pedometer feature to work), the Nokia 5500 Sport further advanced the transformation of a communication device into a lifestyle-management system.

The Nokia phone also had another feature, a so-called fitness diary that could visualize a user's physical activity on its screen with a series of words like *plans*, *goals*, and *targets*. Within a mere four years, the shift from the phone or wearable as a device to count steps to such a device as a physical and lifestyle fitness-management system seemed to be complete.

But there is a missing part to the story of how portable devices would inspire physically engaged ways of relating to sensing machines. In April 2010, a former editor and writer for *WIRED* magazine named Gary Wolf published a long article in the *New York Times Magazine* that named a new cultural transformation that he then saw taking place as people increasingly were tracking, measuring, sharing, and displaying their daily activities. He called this transformation the *data-driven life*.[16]

Wolf named four key technological developments that were ushering in this new datafied existence: the miniaturization of sensing technologies, the ubiquity of smartphones, the sharing cultures of social media, and, perhaps somewhat facetiously, "the rise of a global superintelligence known as the cloud." He argued that the power of numbers from data made available by sensors in smartphones and then-emerging worn devices called *fitness trackers* were bringing about a new kind of "self-knowledge through numbers." Wolf then made a radical claim. Because of these new technical developments, we were entering a new era in which the understanding of our bodies and ourselves would be shaped by sensors. "Automated sensors do more than give us facts," he argued. "They also remind us that our ordinary behavior contains obscure quantitative signals that can be used to inform our behavior, once we learn to read them."[17]

Wolf was no less than describing the origins of a movement that he and another former *WIRED* editor, Kevin Kelly, had tagged the Quantified Self (figure 8.3)—an attempt to brand the growing phenomenon of self-tracking, wherein people recorded and shared data generated by their patterns of behavior, such as sleeping, exercising, or feeling a certain way. But his argument that the belief in knowing yourself through the objective power of numbers was ushered in with sensors and the cloud was slightly historically myopic.

The concept of numericizing and measuring the physical self to gain knowledge is ages old. People have long kept diaries and journals, but these are personal, close-up, and subjective accounts of their experience. French philosopher and social historian Michel Foucault, who long studied the ways in which people shape their own self-hood, claimed that the ancient Roman philosophers called the Stoics sought to obtain "self-mastery" through self-discipline and would trade detailed lists of banal activities that they involved themselves in, such as exercise routines, for example.[18]

Figure 8.3
The Quantified Self phenomenon: meetup, oxygen measuring, and Fitbit tracking dashboard, 2013.

These lists from the Stoics, however, were not personal reflections. They instead were designed to maintain a sense of an objective and distanced "view from above" in order to stand outside of one's self. In other words, the Stoic philosophers used a simple technology, writing, to record actions in order to distance themselves from those actions. As Wolf pointed out, it is the force of newer technologies like sensors that now create this objective distance. His description of portable sensing, recording, and tracking technologies thus falls into this history. Not surprisingly, Wolf's article featured accounts of then-emerging corporations like Fitbit, which began as a fitness tracking company and, after its acquisition by Google in 2021, is now repositioning itself as a healthcare platform.

But do automated sensors really give us objective "facts" about ourselves? To build his argument, Wolf described the activities of a Canadian professor of mechanical engineering named Ken Fyfe, who, in the 1990s, was one of the early researchers to imagine how sensors and bodies physically exerting themselves might add value to each other. While a professor at the University of Alberta, Fyfe, a runner, spent long hours developing a wearable device using accelerometers that could be attached to a runner's shoe to measure speed and distance. Fyfe was interested in portability. How could a runner know about some critical numbers such as speed and distance during the process of running and not in the cooldown period afterward?

Even though research into *accelerometry*, the study of capturing human movement using accelerometers, had been conducted in the 1970s and patents were already being filed in the mid-1980s to use the sensors as pedometers and as exercise-monitoring devices, Fyfe still detected something missing. The gap lay not only in the sensing hardware that could capture a runner's motion but also, more importantly, in the software. Fyfe therefore drew on his expertise in vibrational analysis and in an arcane area of engineering called *gait analysis*, which studies principles of human locomotion—particularly, how we move our feet when we walk or run—in order to gain more information about the act of running while doing it.

When Fyfe began his research, gait analysis was mainly conducted in the laboratory through expensive 3D camera–based tracking systems, but long hours of engineering hacking and prototyping to create a portable sensing machine attached to the foot eventually paid off. It resulted in both a patent for a motion analysis system and a successful spin-off business called Dynastream, which commercialized the small invention: a foot pod sensor.[19] Dynastream eventually sold five hundred thousand of these sensors to sportswear industry mainstays like Nike, Adidas, Polar, Garmin, and others before it was purchased by Garmin.

Fyfe knew that valuable info runners might want to know, such as velocity, the length of a step (stride length), or how fast the foot would hit the pavement, weren't readily available from the pure electronic signals generated by someone running in his accelerometer-augmented shoes. Features like stride length, angular velocity, foot angle, and normal and tangential acceleration were contained in the signal but not apparent to the human eye. They instead would need to be extracted by mathematical procedures in order to make the information humanly readable and useful. Without the mathematics, the sensor readings would be useless to the runners. The quantified information that the runners would want to see while they moved was not hard, cold facts so much as it was the result of particular technologies working together: sensors and signal conditioning and processing algorithms.

The data generated by Fyfe's foot pod sensors that were marketed to Adidas and Nike and other manufacturers was easily displayed in an obvious place: a smartphone app. Today, these dashboard displays in Fitbit's app or Apple's Health app are well known. They give us immediate feedback through numbers, graphs, and curves. The displays attempt to demonstrate the tight coupling between sweat and pixels: our physical actions, like walking or running, and the dashboard's graphic readout.

In fact, hundreds of thousands of people rely on sensor-based data visualizations on their phones and trackers to understand something "objective" about themselves. Fitbits and Apple Watches seem to shout truths about our well-being, and we optimize our physical response, such as by running faster, pacing ourselves differently when walking, or breathing more steadily to slow down our heart rate, based on what they reveal.

But while we think that the visual readouts of bar charts and spinning dials on the dashboard displays of Fitbits and iPhones speak the truth, their data too is the by-product of mathematics. Gary Wolf also emphasizes this in his data-driven life article when he quotes James Park, the cofounder of Fitbit. To really understand the exercising body, Park claimed that "signal processing and statistical analysis" were the key to turning "messy data from cheap sensors into meaningful information."

Ken Fyfe's work extracting useful features from tiny sensors attached to running shoes demonstrates how this meaningful information is obtained. Fitbits and smartphones never display "raw" data. They show us statistically shaped and manipulated features and derived numbers that don't exist before mathematical models are applied to them. This data that we think represents us as fact is indeed the result of specific ways of measuring and quantifying the world through engineering, mathematics, and statistics.

Gary Wolf ends "The Data-Driven Life" with a strong statement:

> Behind the allure of the quantified self is a guess that many of our problems come from simply lacking the instruments to understand who we are. Our memories are poor; we are subject to a range of biases; we can focus our attention on only one or two things at a time. We don't have a pedometer in our feet, or a breathalyzer in our lungs, or a glucose monitor installed into our veins. We lack both the physical and the mental apparatus to take stock of ourselves. We need help from machines.[20]

Remarkably, this statement sounds like it emerged from the nineteenth-century physiologist Étienne-Jules Marey, whom we met in chapter 1 and who also claimed that the new sensing machines of his time would track down "the imperceptible, the fleeting, the tumultuous and the flashing."[21] But as this strange history and the practice of the interaction between sensing machines and our fitness-driven selves demonstrate, the quantified self, brought about from machines, represents only part of the picture. As Spotify Running showed, our desire to "know ourselves" through numbers depends less on what we see and much more on what we physically sense.

Quantification is there, but perhaps not where we first expect it. This quantified self is almost fictitious, depending wholly on mathematical alchemy: the ways numbers are extracted, filtered, and used to shape how instruments and sensing machines act and react to us. Indeed, in contrast to Wolf's argument that "numbers make problems less resonant emotionally but more tractable intellectually," we might say the opposite. The numbers are transparent, and yet they make possible new situations and experiences that resonate with us. Instead of looking for the truth in the "objective facts" of statistical formulations on the dashboard, why not then pay attention to how sensing machines directly interact with us and the world we move in and run through?

9 Machines Like Us

> The danger of computers becoming like humans is not as great as the danger of humans becoming like computers.
> —Konrad Zuse

Microphones listening from thermostats. Cameras embedded in office furniture, observing. Robotic vacuums scanning and making maps of the living room. Contact-tracing apps transparently syncing up with other mobile phones in the vicinity and communicating their data to governments and public health officials. Hardly a day goes by without the release of another exposé on sensors eavesdropping, or watching us.[1] Indeed, as advances in sensing and machine intelligence leave the laboratory at record speed to join the arsenals of the FAANG companies, it seems that the more connectivity we have in the gone-wireless world of sensor-augmented things, the more these devices are assuming human characteristics.

Much ink has been spilt focusing on the manner in which data-hungry corporate entities are sucking up our data, and sensors are usually the first in line to blame. But do these systems actually know as much as we imagine they do? Is there another side to the statement that "once we searched Google, but now Google searches us?"[2] Do we desire interaction with these sensing machines because they seem more like us than first meets the eye? Or, despite the fact that these systems "sense" and "interpret" the world, can we really see them as sensing entities like us?

How much different are our senses, or those of biological creatures, than the senses of machines? It is generally accepted by researchers that the sensory organs, the "sensors" in an organism, determine the way in which the organism perceives its world.[3] From a strictly functional point of view, sensors detect changes in an entity's environment they are monitoring and then take appropriate action with their *affectors*, their motor abilities, based on those changes. The critical piece of information here is that the entity can respond in a specific or distinct manner to the stimuli it senses.

Take the example of the *E. coli* bacterium, a well-studied single-cell entity that has a surprisingly sophisticated sensorimotor system. Using a group of molecules on its outer membrane, the common *E. coli* bacteria are able to sense chemical changes in their environment and correspond accordingly by physically moving toward some chemicals and away from others in their search for food, in a process known as *chemotaxis*.[4] The survival of the bacteria is based on their motility: they either sense food or try to avoid something that would harm them. The bacteria's sensors—in this case, the molecular structures on their exteriors—are therefore critical to their behavior because they enable the bacteria to act in a manner appropriate to the situations they find themselves in at any given moment.

Unfortunately, bacteria have something of a disadvantage. Unlike more highly evolved animals, they have no brain or nervous system. In more developed creatures, nervous systems—sensors, effectors, neurons, and brains—have evolved to coordinate and orchestrate their behavior, particularly developing ways to mediate between perception and action. They choose the behaviors that are most appropriate given the current sensed scene—for instance, realizing organismic goals such as maintaining homeostasis, oxygen supply, water balance, nutritional state, and protection from predators and finding mates. There is an ongoing cycle of perception and coordination in pursuit of the current most pressing goals, together with actions that try and fulfill them.

Other interesting examples are cephalopods, like the octopus. Octopuses are a wonder of sensory and neural technologies. They can sense light without seeing, mainly from light-sensitive proteins found in the surface of their skin. The octopus's tentacles are also filled with sensors—some ten thousand touch and taste neurons. Even more astonishingly, it seems that the octopus's brain and its five hundred million neurons are somehow "severed" from each other and diffused across the whole entity. As one researcher wrote, "It is not clear where the brain itself begins and ends."[5] Despite this neuronal distribution, the creature is still able to coordinate sensorimotor action across its entire body.

Sensory organs also enable action on the part of an entity for it to realize embedded goals. Early on, as hunter-gatherers, we humans needed to survive, and so our senses were attuned to changes in the environment that would enable us to fulfill the goal of survival. We needed to know what something was that approached us and where it was. In other words, sensors in biological entities have evolved to be part of a larger goal-seeking-based form of organization that comprises complex feedback loops among the neural system, the brain, and the environment and that constantly reacts and responds to changes in that environment to sustain the sensing entity.

But what our sensors and effectors mainly do is help mediate between our internal states and the world outside of us. Sensory systems help reduce *uncertainty* in the external world for the organism, which thus allows us to take action. While our internal senses (the stretch of organs, thirst, hunger, or emotional states) give us a readout of own internal states, external sensory systems (the sense organs) reach out to interact with the external environment. These sensors sense—that is, *measure*—the current state of the environment and attempt to make distinctions about particular conditions in it. For example, is a certain chemical present in the environment or not? Does an object move or not?

The process of sensing/measuring thus involves making *distinctions*, choosing one outcome from among many possible ones, and then relating those distinctions to our nervous systems and our effectors, the motor system.[6] The effectors then directly act on those states by mapping them to specific motor actions. The goal of this process is clear: by acting on the world, we attempt to change it in some way. This process of reading one definite outcome from many possible ones provides what researchers specifically call *information* (a measure of possible choices) to the organism, related to the sensor-environment interaction, which then reduces the uncertainty.[7]

To put this in a slightly simpler way, sensors and the senses establish boundaries between our bodies and selves and the environment outside of us. They articulate *differences* between potential states in the environment and convert these environmental states and conditions into specifically distinct neural signals that the brain and the body can then act on. The structures of our sensors and actuators thus allow us to "sense some perceived state of the world, to make a prediction for what is the best action to take given that perception, and to act accordingly."[8] One could say that with respect to some goal or purpose, sensing, in the words of anthropologist and cybernetician Gregory Bateson, "is the difference that makes a difference."[9]

~~

Biological entities or agents are not the only ones to have goals or purposes. Machines can also utilize their sensors to understand something about the environment they are embedded in and then act on it. In fact, one of the things that links humans and machines is that both can be functionally organized for goal-driven, purposive-adaptive perception-coordination-action.[10] The difference is that our goals have evolved based on long-term interaction with the environment. Machines' goals have been established by their designers.

There is a technical term for this ability for a machine to sense and act on its environment: *machine perception*. Machine perception is "the capability of a computer system to interpret data in a manner that is similar to the way humans use their senses to relate to the world around them."[11] Including a wide range of research areas, such as computer-based vision, hearing, touch, smell, and taste, as well as speech recognition, machine perception aims to give machines sensorimotor characteristics, mimicking human perceptual capabilities within a technical system.

But there is a slight problem. If one of the core elements of sensation and perception is indeed how the structure of a stimulus in the external world is measured, imported into the neural realm, and turned into meaningful patterns for the neural-effector system, then what is such a pattern to a machine that has no real nervous system?

To answer this, we first need to understand how a machine sensing system actually detects patterns. A *pattern* is a particular form, organization, or configuration of distinctions or differences—for example, a rhythm, or a geometric figure, like a triangle. In the late 1950s, researchers became interested in teaching machines to recognize shapes, read handwriting, and perform similar tasks through a process known as *pattern recognition* or *pattern classification*. Human and machine pattern recognition both have similar aims: to recognize the presence of a pattern in a given stimulus.

Pattern recognition in machines is based on measuring their surroundings through sensors, analyzing that sensory data, and reacting appropriately to certain relevant events, like a reappearing object or sequence. The task pattern recognition sets out to perform is to discern, "learn," and classify those patterns—to distinguish the signals from the noise. This process involves *recognition*—that is, classifying the similarity of a given stimulus to other previously presented or perceived stimuli.[12]

At first, giving machines the ability to recognize patterns sounds like the domain of so-called artificial intelligence. During the same period in which pattern recognition emerged, however, scientists and mathematicians working with early digital computers sought to define artificial or machine intelligence from a different direction: less as the detection of patterns and more as symbolic operations on logic-driven propositions.[13]

Computer scientist and author Douglas Hofstadter, whose bestselling book *Gödel, Escher, Bach* also focused on mathematics, symmetry, and patterns, described this difference between early AI and pattern recognition:

> In the attempts to get machines to do things such as recognize handwriting or sequences of shapes . . . researchers were faced with questions like . . . "What is the essence of dog-ness or house-ness? . . . What is the essence of a given person's face, that it will not be confused with other people's faces? . . . How to convey these things to computers, which seem to be best at dealing with hard-edged categories—categories having crystal-clear, perfectly sharp boundaries? These kinds of perceptual challenges, despite their formidable, bristling difficulties, were at one time viewed by most members of the AI community as a low-level obstacle to be overcome en route to intelligence—most as a nuisance that they would have liked to, but couldn't quite, ignore.[14]

As Hofstadter makes apparent, here were two radically different worldviews about intelligence: one based on *perception*, the ability to make sense of the world through sensors and effectors, and the other based on *reasoning*, the ability to logically think, construct, and organize concepts.

Yet the introduction of sensory capabilities to machines introduced a new set of challenges to pattern detection that was not just about the differentiation between objects—one of the major social-political concerns around certain techniques in machine learning in which human beings are wrongly differentiated or classified by algorithms that learn patterns from limited and biased data sets. They also introduced *time* into the process of perception. For example, sensors have what physicists call *time constants* or response times. An accelerometer delivers a very fast response because it measures the vibration of an object, which can change rapidly. If an object starts to vibrate or move back and forth in a periodic fashion, we might say that it quickly varies over time.

Take another example. If you are wearing an accelerometer and you jump up and down in a steady, periodic (i.e., recurring at regular intervals) manner, the pattern picked up from the device will be regular. On a graph that shows the relationship between the intensity (amplitude) of the signal and time, you can detect regularly shaped curves. If you stay relatively still, the data will be mostly flat. If you then jerk your body while jumping up and down, however, the pattern will suddenly change, becoming at times irregular or producing sudden bursts of activity—what are called *peaks* (see figure 9.1). These variances can thus be recorded by the sensor and indexed as points in time—what is called a *time series* in statistics.

The amount of variation in a signal we record in the time series is also based on how much detail we can actually zoom in on—something in a digital machine that is dependent on factors like sample resolution, bit rate, and available memory to store long sequences of data. For instance, if you

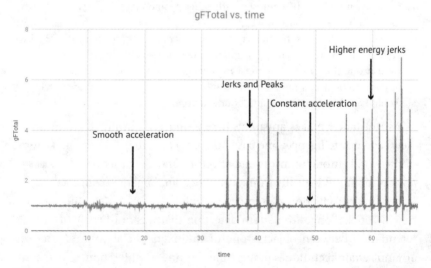

Figure 9.1
The behavior and patterns of an accelerometer over time.

wear a fitness tracker and record data from the device's accelerometer over a twenty-four-hour period, when you look at the overall signals in the time series you will see periods of regular, vibrant activity contrasted with little change. But if you zero in on a shorter time window (say, a few minutes while running), you may see a much greater variation.

Contrast this with sensors that measure signals that change slowly, such as humidity in an environment or the slow shifting of light from late afternoon to dusk, for which we will witness a much slower time evolution. If we have software that allows us to visually zoom in on the signal and observe a few seconds of data, we'll see that there is little change over a short time window (outside of the inevitable noise that we try and filter out) and that only by looking at the entire signal over a longer duration (for light changes, more than a few hours) do we perceive any dramatic variation. What you see (in the displayed signal) is *not* necessarily what you get. In other words, it's all a matter of time: where and *when* you choose to look in the signal.

The other key idea about pattern classification is that it can be highly dependent on guesswork; it is more of an art than a science. While there are dozens of mathematical techniques deployed by software to try and make educated guesses about what will come next in a pattern, much of this detection and, ultimately, prediction runs on a trial-and-error basis. Will a signal that has changed little in the past remain static in the future, or will it radically change? How far ahead into the future can the algorithm predict what the pattern will be? To guess what might happen next in a sequence, machine perception thus employs an *engineering-based understanding of perception*: classifying and making inferences from streams of sensor data using a thicket of mathematical and statistical processes.

What is the process for machines to sense and make sense of incoming stimuli? Machine perception generally can be boiled down to four major areas: sensing, preprocessing, feature extraction, and classification or prediction.[15]

Because the external world is noisy, the signals that sensors pick up usually need to be preprocessed in order to eliminate the extraneous and the unwanted through techniques such as *segmentation* (cutting up or isolating data into smaller sections), *filtering* (removing artifacts or noise), or *compression* (reducing data).

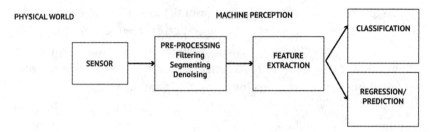

Figure 9.2
The pattern classification process in machine perception.

Once the signal bypasses this initial step, however, the real perceptual transformation begins. Because electronic devices don't intrinsically "know" what they are sensing within an environment, a whole series of signal-processing techniques for the sensors to "make sense" have to be harnessed. These techniques attempt to understand what has happened and what might happen in the future—in other words, prediction. Usually, this process of data mining starts at the level of *features*—the unique characteristics that one wants to isolate from a stream of data, such as the *frequency* (the number of times something is repeated within a given window of time) of a particular movement; a specific shape formed by pixel groupings in an image captured by a live camera; or the structure of a series of vowels in speech.

Feature extraction is a scientific construct. This was acutely demonstrated in 1959 by two neuroscientists, named David Hubel and Torsten Wiesel. The research that the two conducted, which would later win them the Nobel Prize, showed that specific individual neuronal cells in anaesthetized cats would respond to distinct features in an image, such as edges, while other cells responded to different things, such as the angle of a visual object or a line or curve. In other words, visual perception can be described as a process of relaying images from the retina to the visual cortex, which is equipped with what are called *feature detectors*: groups of neurons that are hierarchically organized to pick up or detect specific features of an image and then can relay these features to different parts of the visual system.[16] Visual images can be broken down into distinct subcomponents, each of which is analyzed by a particular set of neurons dedicated to detecting those subcomponents. An image can thus be decomposed into different structures, such as individual edges, lines, or higher-level shapes.

Whether such feature detection is actually taking place in the human brain is a long-running scientific debate. But even if feature detection is not actually an operating principle in neural systems, it hasn't deterred computer scientists from deploying the technique for machines to distinguish and categorize patterns in images, sound, or speech.

Computer vision based systems that seek out distinct characteristics in an image to analyze it and try and recognize shapes in a field of random pixels form one area in which feature detection reigns supreme. The technique is also a central paradigm used in deep learning–based artificial intelligence, including one particularly successful algorithm called a *convolutional neural network* (CNN), which breaks down an image into individual features to distinguish one image from another.[17] In fact, the CNN, which is widely used

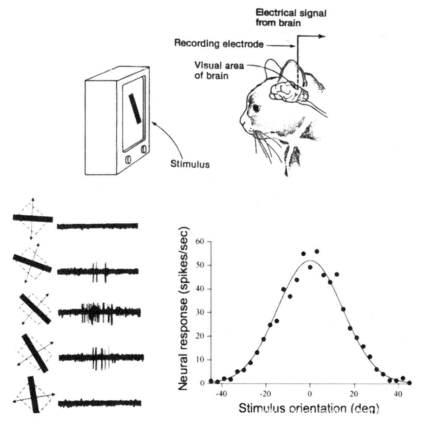

Figure 9.3
Hubel and Wiesel feature detectors.

by Google and other companies for its rapid image-detection capabilities, is directly based on Hubel and Wiesel's model of brain-level feature detectors and the hierarchical organization of neurons that are claimed to be able to recognize, that is, classify or label images. The CNN has had some major successes, for example, in its ability to discriminate between different types of cancer tumors in medical scans. It has also been successful at discriminating against something else: faces that are not white, making its mass deployment more than ethically problematic.[18]

Feature detection has become the holy grail of machine sensing. From an engineering perspective, the technique is useful because it enables reductions to take place in a data set, which makes it more probable to examine the "correct" data. But in addition to the ethical issues around fairness and accountability in these models, the entire concept of "features" also presupposes a certain philosophical view of the world—a relatively static one in which there are individual, atomistic "pixels" or discrete elements of things called features "out there" that the visual perception system seeks to passively "pick up" or recognize.[19]

Machine sensing has functional organizational similarities then that are not too far from us—not least in the fact that machines have the ability, through mathematical and statistical techniques, to *sense* (extract relevant stimuli from the environment), *have sensations* (cause certain sets of processes to take place as the things they sense are internalized in the computer), *detect, group* and *classify patterns*, and, finally, *take some action* that changes the environment they are in.

But there are some core differences between human and machine perception. Despite the use of artificial neural networks, such models don't come close to the sophistication of biological nervous systems. As far as we know, machines don't have nervous systems; they are still nonconscious, although we don't really know what the *umwelt*, the immediate environment that these systems really sense, is.[20] They don't process their sensations by converting stimuli into sequences of neural spikes that a brain then classifies or groups to make sense of. While machines can have goals, these are designed and cannot be reorganized by the machine's internal structure. Unlike brains and nervous systems, machines have less massive parallel and interconnected processing. Finally, unlike biological systems, which can improvise or mutate, machines are fundamentally deterministic—that is, predictable in that they will always produce the same output and their state at any point in the future can be determined from their initial state.

There are other major differences. Artificial electronic or mechanical sensors are singular, isolated units, cut off from the more holistic connections to other parts of the neuromuscular-skeletal organ system that biological sensing systems have. While there are sensors that move (e.g., cameras attached to robotic heads or arms) or that send out information into the environment and await its return (e.g., sonar or ultrasound), unlike biological sensing systems that actively seek out information in the environment in order to adapt and sustain the "organism" they are attached to, artificial sensors are mostly passive; they have little motility. Sensors as isolated units also have difficulty separating desired signals from noise. If a human sensory system encounters noise in one sense modality, it utilizes its other senses to overcome the disturbances. Artificial sensors are not so capable, as engineers will be the first to admit.

But if patterns in biological systems can also be considered informational structures that are picked up or detected by the sensory organs, transmitted to the brain, and used to affect the behavior of a system and enable it to adapt to and survive in a continually changing environment, then patterns in artificial sensing systems might also change the behavior of an artificial system. After all, according to mathematician and cybernetics founder Norbert Wiener, a pattern is simply a form of organization that makes up the fundamental structure of systems, regardless of whether that system is an electrical signal network or a cellular system. It ultimately doesn't matter, as he famously stated, whether the entity is chemical or computational; it will always be changing, producing ever-new temporal behaviors and structures. As Wiener made clear, "We are but whirlpools in a river of ever-flowing water. We are not stuff that abides, but patterns that perpetuate themselves."[21]

The engineers and designers of contemporary sensing machines seek a central goal: to add perceptual capabilities to their systems, much like biological sensing systems possess. But a direct, one-to-one comparison between human (or, more generally, biological) sensing and machine sensing is more complex than it first appears. We may, in fact, be assigning characteristics (or blame) to sensors that are not wholly dependent on the sensor itself but on a larger social-technical combination of factors that the sensor is entangled within.

For example, the quickly growing field of machine listening researches how audio signals that are detected and processed through the interplay of stimuli, ear, and brain in humans can be modeled in computational processes. Features of human hearing such as pitch detection, the ability to locate the position of sounds in space, the perception of loudness, and tone color or *timbre* are key aspects of human hearing that we would like to replicate in machines.

But what parts of human hearing are useful for machines? There is a difference between recording and analyzing an audio signal with a microphone versus making meaning out of that signal. As one expert in human and machine hearing at Google writes, "Our computers are presently mostly deaf, in that they have little idea what the sounds they store and serve represent."[22]

Machine listening is similar to human listening if we see both based on the same principle: the goal and attention-driven perception of sound. This goal can already be seen in devices that have gained a certain level of notoriety at present—Amazon and Google's voice-based services, such as Alexa or the Google Assistant, and the question of whether or not they actually are "eavesdropping" on our conversations.

When you speak, the Amazon Echo device that hosts the distinctly gendered and yet disembodied Alexa is "passively listening" with the assistance of a sophisticated array of seven tiny microphones—a technological breakthrough in a research area known as *sound source localization* that enables the device to function in messy, noisy, real-world contexts, just like our ears do. Using a signal processing technique called *beamforming*, a group of microphones can extract and enhance a desired signal from a specific direction while reducing the interference from undesired signals coming from other directions.[23]

Passive listening means that Alexa is recording a small segment of ambient audio into a buffer and continually writing over that buffer until it detects a so-called *wakeword*: a trigger word that the user inputs (usually *Alexa*) to kick the device into an active listening and recording mode. It's only once the device detects the wakeword that the Echo begins recording and streaming the audio to Amazon's cloud servers, which await with automatic speech recognition (ASR), text-to-speech, and natural language processing (NLP) techniques to make some sense of what has been said and then send this back to the device for response.[24]

The wakeword is a keyword that the user directly tells the system to listen for in order to bring it into active listening mode. But could Amazon also be listening for other words as well? Much media attention has been given over to an emerging technology known as *voice sniffing*. In fact, in 2015, Amazon did indeed receive a patent for "Keyword Determinations from Voice Data." Here, a so-called voice sniffer algorithm aims to "extract keywords, phrases, or other information that is relevant to the user speaking the content" and to transfer those keywords or phrases so that "a recommendation engine executing on one or more of the application servers of the content provider can receive a request to serve a particular type of content (e.g., advertising) to a user, and can determine keywords associated with that user using information stored in the user and/or keyword data stores."[25]

Unlike the singular wakeword model, if specific words or phrases are spoken, the device can kick more quickly into an active listening mode, recording the adjacent sentences around the wakeword and placing these into a buffer, where the file could be subjected to detailed audio analysis: the detection of associated keywords or phrases (context), for example. Finally, the audio data could be further evaluated using a potential range of different speech recognition or natural language processing algorithms—including a well-known mathematical procedure, a hidden Markov model (HMM), which characterizes the statistical properties of a signal by analyzing the most likely pattern or sequence of spoken words in relationship to the actual audio signal.[26] In other words, as you can tell from this technically dry description, the voice sniffer has a clear goal: to recognize features in an audio segment by trying to extract specific information from the signal and, based on this, predict what might be said next.

But goal-directed perception is only one part of listening. We as humans also move from passive to active mode in listening, for example, shifting our attention when we are walking and someone calls out our name. Listening doesn't only involve brute information retrieval and analysis, like machines execute. Listening is also embedded in sociocultural contexts; it doesn't happen just in our ears but involves our other senses, like vision, as well. Music, sound, and speech trigger associations, memories, connections.

Listening produces and enables meaning for us such that when we hear things, we make connections and take actions. While machines like the Echo excel at specific duties, like responding to distinct keywords, they aren't designed to create connections to or associations with things beyond their limited task set.

Instead, it might be easier to claim that the real listener or eavesdropper is Amazon itself rather than these sensing machines that bear the brunt of our privacy concerns and conspiracy projections. Voice analyzers and "filters" are not just in Amazon's sensors and supposedly immaterial clouds. They are also real material in the form of low-paid human laborers, listening in Romania, India, and the United States to thousands of hours of recorded speech from users and attempting to detect and annotate keywords that can be used to retrain algorithms.[27] Thus, while the engineers who are paid six-figure-and-above salaries by Amazon design the next generation of automated machines, the real laborers who enable our supposedly AI-based convenience are toiling away across the world in most likely less desirable conditions.

On October 29, 2018, another example of intelligent sensors with perceptual capabilities supposedly better than those of humans received worldwide attention. The Indonesian airline Lion Air's flight 610 plunged into the Java Sea shortly after takeoff, killing all 189 passengers and crew on board. Five months later, the situation was repeated in another country: this time Ethiopian Air flight 302 plummeted straight into the ground of Addis Ababa six minutes after takeoff, killing 157 passengers and crew.

Boeing, the Seattle-based airline manufacturer that designed and built the 737 MAX planes involved in this set of tragedies, was not the only entity to receive worldwide attention. Measurement devices mounted on these planes, called *angle-of-attack sensors*, which indicate to the pilots whether a plane has enough lift to hold it in the air, were also given prominent media coverage. It seemed that these sensors, which are mounted on each side of the cockpit and designed to measure the angle between the wing and the oncoming air so as to prevent an aircraft from stalling if its angle becomes too steep, had grossly malfunctioned. The sensors indicated a reality different than what the planes themselves were experiencing.

This was not all. Because sensors are basically useless electronic components without their software "nervous systems," the angle-of-attack measurements produced a heavy-duty response from the Maneuvering Characteristics Augmentation System (MCAS) software control, causing stall warnings to be announced and the aircraft to suddenly force the nose down so as to prevent it from literally stopping in the sky.[28] The 737 MAX's "erroneous sensor inputs" might thus have been responsible for causing the planes to nosedive from the sky. Unbelievably, eight days after the Lion Air crash, Boeing issued a bulletin on its intranet site with the technically euphemistic headline "Uncommanded Nose Down Stabilizer Trim Due to Erroneous Angle of Attack (AOA) during Manual Flight Only."[29] In other words, Boeing seemed to cautiously admit that the cybernetic loop of regulating feedback between the human pilot and the machine had failed, sending innocent passengers to a terrifying death.

Fingers point at multiple causes for the disasters: automation that increasingly removed human control from the process of flying, coupled with bad design decisions from Boeing's human designers and board of directors, who cut as many corners as possible to maximize its profit margins with the sale of the planes. During its development, the MCAS increasingly suffered what is known in engineering as *feature bloat*. The system became ever more automated and its operations increasingly more hidden. Boeing not only concealed the existence of the MCAS system from its pilots. It also charged an extra fee for its safety features: a "disagree light" that indicated if the two angle-of-attack sensors generated contradictory measurements between each other and an "indicator" readout that depicted how much the plane's nose actually tilted, both of which could be used by human pilots to disarm the automated software system. It seemed that "human error" amplified by technological failure was the real culprit in the collapse of not only a supposedly innovative guidance system but also Boeing's attempt to build a new flying machine for the twenty-first century.

What then does this preventable calamity prove? That despite the belief that sensors are seen as all-knowing and all-observing devices, in the midst of malfunctioning technologies that cause airplanes to nosedive, personal assistants that listen to what we say, and robotic vacuums that map our living rooms, it is still human beings who lurk in the background of our allegedly autonomous technologies. Although our machines are increasingly given new possibilities of percepts and senses, the increasing chance to understand patterns and to thus make sense of the world for themselves, the real ghost in the machine still seems to be us.

10 Becoming Aware

> The machine gives our dreams their audacity: they can be realized.
> —Le Corbusier

In 1959, Dutch artist Constant Anton Nieuwenhuys authored a short text that was published in *Potlatch*, an internal newsletter of the left-wing Situationist International movement. "The technical inventions that humanity has at its disposal today will play a major role in the construction of the ambiance-cities of the future," he wrote. "It is worth noting that significantly, to date, these inventions have in no way contributed to existing cultural activities and that creative artists have not known what to do with them."[1]

Nieuwenhuys didn't have at his disposal the sensing machines we have today to make his vision of a future "ambient" city viable. Nevertheless, the essay he penned, titled "The Great Game to Come," set out an extraordinary vision: a dynamic, ever-shifting urban space that would counter the standardized dullness of postwar urban life. The artist attempted to paint a vivid portrait of such an ambient city, where "static, unchanging elements must be avoided" in favor of spaces that would be in constant motion, rapidly changing their appearance. The goal of such a city was unmistakable—to use new technologies to introduce play into the fabric of urban social life.

Nieuwenhuys's dream of the playful city was inspired by a fellow Dutch compatriot, the historian Johann von Huizinga, whose 1938 book *Homo Ludens* ("Man the Player") passionately argued that play, whether in animals or humans, is an essential part of biological life.[2] Nieuwenhuys was so inspired by Huizinga's argument in fact, that he began work on one of the greatest unrealized utopian urban schemes of the twentieth century: a new city that he appropriately named New Babylon.

New Babylon didn't consist of concrete and steel, glass, or wood like other urban environments but was constructed of models, maps, films, collaged images, and words. Its inhabitants would involve themselves in an "uninterrupted process of creation and re-creation" in which automation would replace the tedium of work with "a nomadic life of creative play."[3] Technology would be the central factor in establishing the social organization of the city—a hierarchy consisting of "sectors" and "networks" that would connect all inhabitants into a dynamic set of relations with their surroundings.

As a new kind of city, New Babylon's complex spatial organization, one of interwoven levels of transport networks and artificial and natural spaces all linked together by communications infrastructures, defied "traditional means of cartography." Instead, to understand and manage the flows between different layers of the city's infrastructural networks, "recourse to a computer will doubtless be necessary to resolve such a complex problem."[4]

Nieuwenhuys's pronouncement seems prophetic, and not only in the suggestion that computers would be needed to coordinate so much overlapping and interwoven activity. No doubt computers would have also been marshaled to achieve the vision of an environment in which "each person can at any moment, in any place, alter the ambiance by adjusting the sound volume, the brightness of the light, the olfactive ambiance or the temperature."[5] In other words, spaces would have to somehow be "aware" of the activities taking place within them so that the environment could know when to change its appearance and behavior.

Nieuwenhuys conceived and worked on the concept of New Babylon between 1956 and 1974. Like many utopian urban projects, it was never built. But little did the artist know that the technologies to enable his new vision of dynamic and aware spaces were actually being invented and deployed on the other side of the Atlantic for a different and less utopian purpose.

In the late 1950s, the Western Electric and the American Telephone and Telegraph corporations developed the first distributed sensor system through the support of the Office of Naval Research (ONR), the largest Cold War military defense technology funder. Initially prototyped off the coast of New Jersey and in the Bahamas in the Atlantic Ocean, the system was a classified project for the US Navy named Sound Surveillance System or SOSUS.[6]

SOSUS was an underwater tracking system consisting of distributed arrays of acoustic sensors called *hydrophones*, which were positioned on the ocean floor to monitor moving acoustic signals generated by Soviet submarines. Researchers had learned that it was possible for acoustic signals to reflect off the bottom of the ocean floor and surface as they traveled through deep canyon depressions over thousands of miles with little interruption. Around the same period, improvements by the US Navy to deep-water temperature sensors called *bathythermographs* that could operate at depths of over three hundred meters also enabled scientists to measure the changing propagation speed of sound waves under the water, particularly their difference in speed based on variations in water temperature.

The SOSUS system could thus detect long range sound propagation through a network of distributed sensors. The hydrophones, installed in one-thousand-foot-long sequences called *arrays* on the ocean floor, could discern the angle and direction of a sound traveling hundreds of miles and then track these sound waves as they moved across the ocean, with each hydrophone forming a network of signal-chasing devices. The signals were then sent to massive machines on land, where they were analyzed using the new speech synthesis technologies developed by AT&T's Bell Labs—most specifically, the sound spectrograph, which visualized acoustic frequencies and their amplitudes over time and which was modified to analyze the low-frequency sounds produced by submarines. These machines enabled so-called low-frequency analysis and recording (LOFAR) to be conducted on the signals in real time, thus separating and differentiating a submarine's low-frequency machinic sounds from the general noise of the ocean. SOSUS was so successful that it was eventually deployed in both the Atlantic and Pacific Oceans and, in particular, the vulnerable waters around the Greenland, Iceland and the UK (GIUK) Gap, where Soviet subs would enter the North Atlantic.

SOSUS is said to be the first system to position sensors across distance in the effort to monitor a large-scale area, but another fabled military project deploying distributed sensors has a different story to tell. During the Vietnam War, US military forces faced a central conundrum. Despite continued bombings and massive reinforcements from the US and South Vietnamese armies, the North Vietnamese would continually succeed in smuggling arms and equipment to their counterparts in the south of the country. Part of this could be attributed to the unforgiving density of the Vietnamese jungle, which the North Vietnamese intimately knew, as well as "the complex network of roads, trails and footpaths, the constantly changing weather conditions, and the rugged terrain" that made "the detection of this infiltrating traffic difficult."[7]

While US Defense Secretary and Vietnam War architect Robert McNamara demanded the building of solid walls to cut off the enemy and their supply routes (interestingly calling to mind the border wall visions of a later US president), a secretly funded think tank of physicists and scientists working for the Advanced Research Projects Agency (ARPA; later adding "Defense" to the name to become DARPA), called "Jasons," suggested something more cutting-edge: an "invisible electronic fence" that eventually would form an early prototype for so-called smart warfare and that was astonishingly rethought later to be deployed in US/Mexico border security contexts.[8]

The US Department of Defense (DOD) set out to fund and construct what it called an *air support anti-infiltration barrier* for gathering intelligence—what eventually became known as *McNamara's line* or *McNamara's wall*. To achieve this virtual fence, some twenty thousand wireless sensors were dropped from the air or planted on the ground in Laos and Vietnam between 1967 and 1972. Many of these battery-powered technologies were more than exotic, such as bio "people sniffers" designed to detect the scent of urine, body sweat, or lingering traces of motor exhaust. Others drew on more proven seismic and acoustic technologies—some, in fact, based on the underwater hydrophones used in the SOSUS system.[9]

In projects that would eventually go by the code names Igloo White and Practice Nine, the operation of the sensors was similar to the SOSUS system: tracking the flow of ground vibrations or sounds, presumably produced by vehicles as they moved from one position to the next. Seismic sensors called *spike buoys* or *seismic intrusion detectors* (SIDs) could detect ground motion triggered by trucks as off-site teams performed real-time spectral analysis on the acoustic signals to distinguish them from the sounds of birds or other animals, thus reducing the possibility of bombing these creatures. The network of sensors would then be activated in a chain-reaction-like sequence and then, via radio frequency, transmit their locations to overhead reconnaissance flights, which consequently would request fire support for bombing raids (figure 10.1, top).[10]

This fantasy of a sensor-laden battlefield was predestined to become a future model of war. The main US commander of forces in Vietnam, General William Westmoreland, implied as much when he stated, "I see battlefields under 24-hour real or near-real time surveillance of all types."[11] To encourage Westmoreland's vision, in the 1980s DARPA thus began exploring a new kind of combat potential dubbed *network-centric warfare*. According to the DOD, the sensor-oriented battlefield would involve "an information superiority-enabled concept of operations that generates increased combat power by networking sensors, decision makers, and shooters to achieve shared awareness, increased speed of command, higher tempo of operations, greater lethality, increased survivability, and a degree of self-synchronization."[12]

To enable this ambitious plan for "greater lethality," DARPA launched the Distributed Sensor Networks project in 1980—a multi-billion-dollar research program enrolling universities such as MIT and Carnegie Mellon to research and develop networks of spatially distributed ground-based sensors for aerial "distributed target surveillance and tracking" (figure 10.1, middle).[13] The project concluded at the end of the 1980s, but DARPA didn't stop there. A series of similar sensor research programs were funded and carried through the 1990s and 2000s, with names like SensIT and SIGMA+.

In the twenty-first century, military and civilian research into such distributed sensor technologies continues apace, but the real cutting edge, at least for military use, goes far beyond electronics. The new sensors are organic—namely, plants. DARPAs 2017 founded Advanced Plant Technology (APT) funding program works with, modifies, and genetically reprograms plants to turn them into "organic surveillance sensors" in order to detect "chemical, biological, radiological, and/or nuclear threats, as well as electromagnetic signals" (figure 10.1, bottom).[14] DARPA's simple but perverse aim to create a "new sensing platform that is energy independent, robust, stealthy, and easily distributed" sounds hard to believe, but in fact the agency had already been funding basic research into such organic modifications for military purposes for decades.

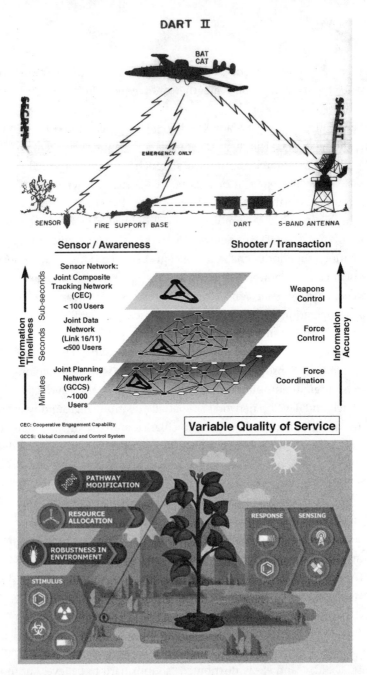

Figure 10.1

Three visions of sensor-based warfare. *Top:* Operation Igloo White illustration (1970), from Project CHECO Report. *Middle:* Network Centric Warfare diagram, "Variation of Quality of Service across the Warfighting Enterprise" (1998). *Bottom:* Infographic from DARPA describing the Advanced Plant Technologies research program (2017).

The sensors of the Sound Surveillance System and Project Igloo are earlier prototypes of a type of sensing machine known today as a distributed, *wireless sensor network*.[15] The basic concept of a wireless sensor network is a spatially distributed group of small, lightweight, and power-efficient sensors and transceivers that can be scaled over large distances in both dense and sparse configurations. These sensors monitor systems or environments around such characteristics as temperature, pressure, humidity, sound, gravity, or motion. Think of any environment—the physical or biological world of a forest, a desert, the floor of a factory, under the sea, or even, as we'll see, on our bodies—and then imagine how to understand what is taking place in such areas, what researchers call the *sensor fields*.

Although they differ in terms of design, spatial layout (what is called *topology*), and manufacturer, most sensor networks have a more or less similar hardware and software architecture. First, independent sensor nodes in a wireless network acquire data about something such as air quality, temperature, motion, a change in molecule composition, or noise levels. This is what is called a *signal* in the lingua franca of engineering. After preprocessing and filtering, this data is then wirelessly sent from the localized node using one of a half dozen possible communications protocols.

Because it has been reduced, the data takes up lower bandwidth, allowing it to reach a mid-point base-station called a *sink* or even an endpoint computer in a faster, more efficient manner. As sensors in a wireless network are spatially distributed, they normally ask or *query* each other to find out what their neighbors are doing. Through their software, they may also reorganize their connections in an ad hoc manner. Now imagine this process happening both synchronously and asynchronously with hundreds or even thousands of nodes and you have the essence of a wireless sensor network.

Much of the research in these sensor networks is in rather cryptic engineering concepts, such as the development of new embedded operating systems that run on smaller microprocessors, communication protocols to send signals over long distances, energy saving and harvesting techniques, data traffic management, and performance measurement and benchmarking.

Such sensor arrays are also deployed in a vast myriad of contexts. In fact, one standard reference textbook lists no less than two hundred applications, ranging from the everyday, such as process control and mobile robotics, to the exotic, such as sensor networks for theme parks or tracking smart bullets fired from a paintball gun using wireless transmission and batteries.[16] Despite the small size of the components, such sensor networks are beyond ubiquitous. They now cover almost every square mile of the earth, turning every place, from the floor of the ocean to a crop field, into an electronically sensitive zone. Urban spaces, highways, the ridges of mountain ranges (for seismic detection), hospitals, the insides of factories, and transport and logistic infrastructures are all new application arenas where sensor networks are taking command.

⌒

The development of distributed, wireless networks of sensors suggested something greater than new modes of battle. They introduced a novel understanding of how sensing machines would reinterpret our perception of space and time. Military-styled distributed sensors sniffing for subs and tanks are as far away from Constant Nieuwenhuys's progressive, humanist vision of technology in the service of a new play-driven society as one can imagine. And yet, it would be through these same military-birthed technologies that scientists, architects, urban planners, and even artists would attempt to realize Nieuwenhuys's dream, that "the variable or changing character of architectural elements is the precondition for a flexible relationship with the events that will take place within them," some fifty years later.[17]

By the mid-1990s, countless research projects from the likes of Intel, the EU, UCLA, UC Berkeley, the Swedish-based Interactive Institute, and others were afoot, with civilian applications for sensor networks emerging in areas like forestry, habitat and climate modeling, seismology, robotics, building information modeling, and even games and art projects. With names that aimed to suggest parallels between an emerging new age of continual but hidden monitoring, systems utilizing dozens if not hundreds of coin-sized or smaller sensors increasingly appeared to be tagged with the moniker of *smart*: Smart-Its (EU), Smart Dust (Berkeley), and SmartMesh.[18]

Alongside these tiny technologies came a host of new paradigms to describe such "smart" sensing systems: ambient intelligence (AmI), pervasive computing, embedded computing (EmNets),[19] sentient computing, and everyware.[20] Other descriptions like situational, pervasive, ubiquitous, and ambient, or newer expressions like fog, edge, or mist computing, have quickly followed. While all of these paradigms have slightly different names, they all attempt to describe a somewhat similar idea: that spaces and environments themselves can be made aware, responsive, and perceptive through vast arrays of tiny, distributed sensors networked to each other, monitoring and processing large amounts of data from territories and environments without the need for human perceivers.

Rooms and environments outfitted with sensor networks can now perceive environmental changes—light, temperature, humidity, sound, or motion—that traverse or move over and through a space. This ability reformulates sensor networks as something akin to bodies because, like human or animal sensing systems, they are also aware of the changing environmental conditions that take place around them, measuring, making distinctions, and reacting accordingly to these changes.

Spaces now obtain a kind of "awareness" by way of sensor-based *percepts*, a word originally used in nineteenth-century psychology to denote any object of human perception but which has now been reconfigured in AI-based sensor research to describe the sequence of inputs a sensor sends to any software agent to be processed. Sensor networks can also gather *multimodal* (i.e., many different senses) data, which then has to undergo a process of feature extraction and fusion—similar to the human perceptual system. Finally, some kind of action needs to take place that alters the environment. The operation of a sensor network is thus a similar but grossly simplified model of how our own perceptual systems function.

Comprehending the technical architecture of distributed sensing machines is not particularly easy. Concepts we formally assigned to biological creatures like senses, percepts, awareness, and body are transformed in sensor networks into modular, logical sets of conditions and instructions. For example, while sensor networks and the spaces they are deployed in appear to simulate the processes by which sensory organs, the brain, and the neuromuscular-skeletal system interact with each other, their structure bears little resemblance to the infinitely more complex perception and action-perception-environment feedback loops that humans and other biological systems are afforded simply by virtue of their evolutionary and environmental makeup.

Different models of sensing also suggest different worldviews. For example, even though *remote sensing*—detecting and obtaining information about objects or areas from a distance via aircraft or satellites—uses cutting-edge laser and image-capture sensor technologies to scan and calculate topographies on the earth's surface or to measure the amount of backscattered radiation in order to monitor changes in surface wind, the model of distance-sensing the earth from the sky in actuality reinforces an old stereotype: the classic, all-knowing God's-eye view from above.[21]

In contrast to satellites monitoring the earth, wireless distributed sensor networks suggest another paradigm, one with more atmospheric or field-like qualities. The field-like qualities invoked by the presence of thousands of Internet of Things–enabled sensing devices spread across geographic distance are precisely why newly emerging paradigms with odd names like *fog*, *edge*, or *mist computing* have been developed.[22] These distributed sensor-network paradigms are all of similar flavor but differentiate themselves based on where data is actually processed within large-scale network-based computing infrastructures: at the site of the sensor device itself, in the servers near the sensors, or in the almighty, faraway cloud.

That thousands of sensors continually sense the world around us without needing our help seems hard to fathom. But the worldwide COVID-19 pandemic in 2020 demonstrates how important such distributed sensing systems are, even given their invisibility and geographical spread. In the summer of 2020, Ying Chen, a UK-based climate researcher, published a scientific article with the somewhat startling headline "COVID-19 Pandemic Imperils Weather Forecasting."[23] Chen had studied how decreases in air traffic caused by pandemic-related flight cancelations over significant areas of the planet, particularly the United States, China, and Australia, as well as more remote locations like the Sahara region, led to a surprising deterioration in weather forecasting accuracy.

A little-known US-based meteorological initiative called Aircraft Meteorological Data Relay (AMDAR) gathers wind and temperature data by putting thousands of sensors into the atmosphere via forty different commercial aircraft, generating some seven hundred thousand weather observations daily.[24] With participants ranging from Lufthansa and United Airlines to the United Parcel Service (UPS) and FedEx, 3,500 airplanes use installed sensors to capture wind speed, air temperature and direction, barometric pressure, water vapor density, and turbulence patterns at specifically agreed upon sampling rates so that the data is consistent across flights. The AMDAR system is basically like having a massively distributed sensor network flying around and sampling weather predictors in the inner atmosphere.

According to Chen's report, the reduction of flights between March 2020 and May 2020 as a direct result of the COVID-19 pandemic resulted in an estimated 50–75 percent loss in weather observations. The report revealed what at first might seem somewhat obvious: compared to similar forecasts from periods before 2020, the winter to spring period of 2020 revealed significant drops in the amount of weather data collected from commercial airplanes and, hence, a reduction in forecasting accuracy.

But more interesting is the fallout from such data loss. In addition to less precise forecasting, Chen argued that economic problems due the pandemic could also be exacerbated. For instance, loss of weather forecasting data could lead to imprecise predictions of energy loads on power grids as "people crank up their air conditioning." Chen said, "'If this uncertainty goes over a threshold, it will introduce unstable voltage for the electrical grid.'" This could lead to blackouts, "'the last thing we want to see in this pandemic.'"[25]

The disruption of weather data gathering might count as a negative effect of the COVID-19 pandemic. But another international distributed sensing system, one used for detecting vibrations at multiple locations around the earth, was also affected for a completely different reason. According to a large-scale research study reported in the peer-reviewed journal *Science* in mid-2020, COVID-19 did not only result in a global reduction in air traffic, leading to inaccurate weather forecasting. The pandemic also produced what researchers have labeled an unprecedented "global quieting."[26]

Using networks of 337 professional and "citizen" seismic sensors spread across the globe, the researchers reported drops of almost 50 percent in noise generated by human activities, ranging from the vibrations produced by transport systems, roadways, and industrial machinery to sporting events. The seismic sensor networks allowed researchers to literally "hear" the pandemic, providing them with an acoustic snapshot of how such human noise can interfere with sensitive monitoring systems installed in quiet places that normally aim to detect seismic vibrations from earthquakes. Indeed, the networks of seismic sensors could perceive a kind of traveling "wave of silence" as COVID-19 lockdowns progressed across the planet, from China to Europe and North America. The conclusion of the study? The sensors enabled not only a detection of the change in seismic vibrations but also something more profound: a real-time portrait of the effect of government-mandated mitigation policies on our daily lives.[27]

⌒

One of Constant Nieuwenhuys's key notions in New Babylon was that emerging electronic and televisual technologies could help replace older preexisting experiences of the city with newer ambient ones. The citizens, called New Babylonians, would be in charge of how these ambiances would evolve. Through their movement and presence, they would be the ones to generate new transformations in the environment they inhabited. Sounds and smells could be altered, along with the flow and feeling of space. Through imagined technologies that remained unnamed by Nieuwenhuys while he developed the conceptual work of New Babylon, the particular spatial and temporal conditions would shape how these technologies could detect changes of behavior and alter spaces according to these changes.

Although sensor networks only reached wide-scale usage in the 1990s, the ideas of Nieuwenhuys and other experimental artists and architects seeking to make reactive and aware cities through sensing and computing technologies were already in the air in the 1960s. During the same period in which Nieuwenhuys was conceptually developing New Babylon, other experimental and utopian architecture projects from Coop Himmelb(l)au and Haus-Rucker-Co in Austria, Yona Friedman in France, Archigram, Julia and John Frazier and Future Systems in the UK, Constantinos Doxiadis in Greece, and the Metabolists in Japan, among others, were setting out fantastic visions, some pure paper and some built, of architectures that would respond and react to their surroundings.[28]

With some of these architects, the concept of admittedly crude distributed sensors making spaces "alive" would play an essential role. The once experimental and now mainstream Viennese architecture collective Coop Himmelb(l)au desired spaces that would become like heartbeats and cities that would pneumatically inflate. A 1968 experimental performance called *Harter Raum* (Hard Space) that took place on the outskirts of Vienna also aimed to prove that sensors were not just data-sucking systems. They could also partner with industrial chemicals—namely, explosive TNT—to literally create space. In the *Harter Raum* experiment, the heartbeats of three individuals monitored by crude stethoscope-like amplifiers were used to set off a series of sixty real explosions in a two-kilometer-long field.[29]

In another experiment called *Astroballoon*, which featured an inflatable environment that a visitor could enter, the collective worked with similar stethoscopes to amplify the heartbeats of visitors and use these to affect arrays of small lights installed within the inflatable. As one architectural historian described it, "By integrating medical technologies as part of their installation, Coop Himmelblau effectively turned architecture inwards, into the very interior of the user, at the same time that amplification technologies turned those interiors inside-out. Heartbeats were registered, broadcast, and externalized."[30]

But these experiments involving flooding city streets with soap bubbles or constructing breathable plastic bubbles with lights turned on and off by visitors' heartbeats were not urban scale. They chiefly took the form of concept sketches, manifestos, texts, and small-scale artistic performances and interventions that sought to use the gallery, the exhibition space, and even the street to put forward a new interactive role between the city and its inhabitants enabled by technology—to make the city itself "beat

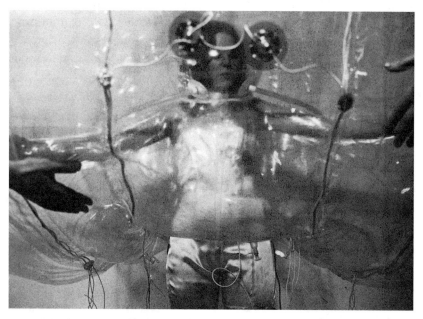

Figure 10.2
Coop Himmelb(l)au, *Astroballoon*, Galerie nächst St. Stephan, Vienna, February 1969. Photo by Erwin Reichman, courtesy COOP HIMMELB(L)AU.

like the heart and fly like breath."[31] Today, architectural theorists give the name *ambient* to these early technologies that Nieuwenhuys and others had imagined or even had the opportunity to toy with. Ambient does not only suggest spatially distributed. It also denotes the ubiquitous, pervasive, and invisible nature of the sensor networks that are enabling ways of thinking about space itself as a responsive and shifting field.

There is a good reason that architects, theorists, and practitioners alike have a keen interest in sensing systems.[32] As New Babylon and other architectural experiments reveal, urbanism and architecture long envisioned such aware environments through new technological interventions. This may be part of the reason that architects, urban planners, policy makers, and corporations like Google's Sidewalk Labs or Cisco Systems have become so obsessed with distributed sensing as a key factor in enabling new smart urban spaces as of late.[33]

Nieuwenhuys didn't have access to the kinds of sensor networks now almost effortlessly deployed to create the sensor city of the future. New Babylon's technological vision was fundamentally another vision of such a smart city—one not marked by large-scale data extraction, environment-killing transport and increasing revenue streams on everything from parking and shopping to health care and utility monitoring. New Babylon was anticapitalist urban growth, a city marked by the belief that new forms of aware technology would somehow, someday release us from the drudgery of labor under advanced capitalism.

The utopian pronouncements of smart cities now go far beyond Nieuwenhuys and Coop Himmelb(l)au's experiments. Smart, responsive cities are now envisioned at scale, made possible by sophisticated sensor-based building management systems collecting and acting on data 24/7.[34] Carlo Ratti, a smart city guru and head of the MIT Senseable City Lab claims that new sensing network paradigms in the smart city are "suffusing every dimension of urban space, transforming it into a computer for living in" (to paraphrase Le Corbusier's machine for living concept from the 1920s). "The rise of ubiquitous computing," according to Ratti, "has created a fundamentally different space—one where digital systems have a very real impact on how we experience, navigate and socialize."[35]

Ubiquitous sensing networks as "a computer to live in" reach their height in the emerging arena of digital building information modeling systems and software modeling techniques of architectural conglomerates such as Arup, the world's largest architectural engineering firm. Founded by Danish-English civil engineer Sir Ove Arup in 1963, the firm initially gained fame for its structural engineering expertise from complex architectural projects, including architect Jørn Utzon's notoriously difficult Sydney Opera House in the 1970s, and it has been responsible for the structural engineering and design architecture of thousands of high-profile projects across the globe since.[36]

The Arup group has expanded into using sensing across all aspects of its architectural and urban planning work, from the integration of real-time data feeds from Internet of Things devices into BIM systems and the utilization of thousands of sensors for infrastructural monitoring on bridges and transport systems to the development of *digital twins*—responsive virtual, 3D computer models that connect physical and digital systems and allow digital features to be altered by real-world sensor data feeds.[37]

Arup's interest in sensing is also apparent in its custom, in-house-developed software systems. One complex data-collection and monitoring system called Global Analyzer allows users to obtain real-time streaming data and visualizations of such data from networks of sensors, measuring geotechnical features like displacement of ground movement, acceleration, wind, and shear velocity to analyze a building site.[38]

An even more totalizing system is the appropriately named Arup Neuron. Neuron is cloud-based "smart building" modeling software that incorporates real-time sensor data and machine learning techniques for energy and building systems optimization. The Neuron system continues the tendency to equate architecture with human bodies, stating in its data sheets that "Neuron is named as a reflection of the human neuron network—just like our own neurons, the IoT sensor network enabled with analytics capabilities is deployed in the building environment, enabling prompt and adaptive response to dynamic environments."[39] More directly, the components of Neuron are compared to human body parts: the brain as the Neuron platform, the blood vessels as the pipes, the bone and skin as the architecture, and the respiratory and circulatory systems as the HVAC and drainage systems.

When it comes to urban infrastructure, however, scale is relative. Singapore, always ahead of the international technology curve, wants to up the stakes, moving past simply creating sensing systems that merely manage smart buildings. The city-state, which vies with Oslo, London, and

Zurich for the most expensive in the world in which to live, now has a new mission: to turn the entire territory into a colossal sensing machine in an effort to become the planet's first "smart nation." Begun in 2014 with the deployment of over a thousand sensors in every conceivable location, from the tops of buildings to street lights running facial-recognition software, Singapore's Everyone, Everything, Everywhere, All the Time (E3A) vision aims to create a central repository for all of this real-time data: a multimillion-dollar Virtual Singapore platform built by French defense contractor Dassault, which will "capture the virtualized life of Singapore."[40] While the director of Singapore's National Research Foundation claims that this new smart nation will "give the right data to the right people at the right level at the right time," there are naturally strong privacy questions concerning this sensor free-for-all, particularly given Singapore's generally lax privacy laws. Indeed, the reality of smart cities is that they are anything but. According to one EU smart city consultant, many initiatives, like Barcelona's widely heralded smart city initiative from the early 2000s, have mainly failed, bringing neither new forms of social communication nor democracy but instead tons of electronic waste and rotting sensing infrastructures.

But resistance does exist against these somewhat banal realizations of Nieuwenhuys's New Babylon. The glaring example of this is Google subsidiary Sidewalk Labs's long-in-the-making showpiece: Toronto's Quayside smart city. Sidewalk Labs sought to turn 190 acres of industrial waterfront real estate into an Innovative Design and Economic Acceleration (IDEA) district—a sensor-fueled "neighborhood of the future," replete with features such as flow-monitor and water-quality sensors for storm water management, sensor-monitored pneumatic trash chutes that could measure the volume and weight of refuse, agricultural sensors to monitor plants and trees in the area, air quality sensors attached to street poles and traffic lights, and sensors for parking spot and traffic management.[41]

Alas, Sidewalk Labs's ambitious plans to reimagine Toronto as a "gigantic data collection machine" were destined not to be.[42] Eventually it was canceled by the city government, not only due to the "unprecedented economic uncertainty from COVID-19" but also because citizens, nonprofit organizations, and advocacy groups relentlessly questioned the creation of such a smart "surveillance city" and the endless data flows produced by its inhabitants that, in essence, would become the property of one of the world's largest corporations.[43]

Architects, urban planners, and artists' visions of spaces becoming bodies and bodies becoming architecture in new kinds of feedback loops between sensors and the environment are materializing today in another way. If twenty-first-century New Babylonian citizens of the new city are the catalysts for change and transformation that Nieuwenhuys originally sought after, then in today's sensor visions these citizen's bodies themselves would also become sources of data.

While wireless sensor networks have successfully turned forests, deserts, the ocean, and now the global city into large, spatially aware, sensed zones, a more recent sensing paradigm called a *body area network* (BAN) or *body sensor network* (BSN) monitors and data collects at a more intimate scale. BANs sprung up in the mid-1990s in an engineering-led effort to reimagine our bodies as a new nexus for information-driven telecommunication. Earlier wireless personal area networks were (and still are) organized around an individual and the proximity of their personal devices instead of transmitting data over a larger area network such as a local area network (LAN) or a wireless local area network (WLAN)—the standard systems that power contemporary electronic telecommunication.

The term *body sensor network* was coined in 2000 by an Imperial College–based researcher to take advantage of the personal notion of transmitting data between devices while adding a new twist: using sensors housed not only on the body but also *inside* it to communicate with and between each other.[44] In other words, a BSN is a complex sensor network embedded on and below your skin. As "a common approach for pervasive monitoring," the network "represents a patient with a number of sensors attached to their body, each sensor also being connected to a small processor, wireless transmitter, and battery, and all together forming a 'BSN node complex' capable of seamlessly integrating with home, office, and hospital environments."[45]

In addition to standard sensors that measure motion, skin temperature, heart rate, skin conductivity, or muscle activities, many of the sensors used in BSNs are also new biosensors that work with chemical or gaseous materials. They combine biological elements (like sweat, saliva, or carbon dioxide) with a physiochemical detector that is often optical or electrochemical. The detector then "interacts with the substance being analyzed, and converts some aspect of it into an electrical signal."[46]

More important is the fact that biosensors in systems like BSNs are only one part of a much larger sensor system.[47] That biosensors can not only be worn on the surface of the skin but also be implanted into us obviously makes them far from noninvasive as many of these sensors function internally: glucose monitors, pH detectors, tissue oxygen sensors, brain stimulators, or even pressure devices for intracranial sensing.

Such profoundly corporeal sensor network infrastructures certainly generate some controversial propositions. Dawn Nafus, an anthropologist who works at Intel as a researcher and has long studied the cultural impact of biosensors using ethnographic methods in observations and interviews with the users of such technologies, demonstrates this complex dynamic when she writes, "The focus on biosensing requires us to slow down our judgement about who is tracking what. It opens up a view onto the diversity of practices possible by making it harder to pretend that we are already know, by looking at the technology itself, what sort of phenomena we are examining."[48]

On the one hand, through their continual monitoring of vital signs, these body-installed networks of sensors not only monitor existing conditions but also aim to be preemptive—heading off health issues before they

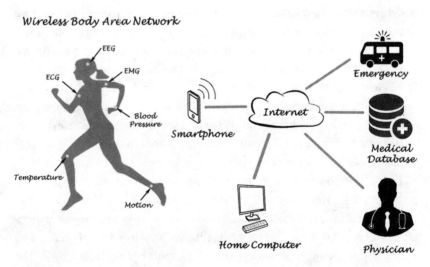

Figure 10.3
Body area network diagram.

actually arise. Preventative health is increasingly dependent on the hardware sensing and software intelligence of such systems, but it is also based on a more complex and nuanced issue: that pervasive, continual monitoring and awareness is a necessary prerequisite for deterring health problems that might or could arise in the future.

At the same time, biosensors networked to each other could eventually "make a home in the body," suggesting that BSNs, in their all-pervasive variety, will somehow become part of us.[49] The potential of such systems is that they not only can be designed for continual monitoring but also can be seen as a persuasive technology as well, alerting us when something seems to go wrong or when something is predicted to go wrong.

The question of prediction leads to an even messier problem. Because networks of sensors know little about the situation they are sensing, they have to be given some form of context. Hardware by itself is basically dumb when it comes to knowing, let alone reflecting on, what it's sensing. That is, even with medical-engineering technology advances, given the sensitive and indeed life-and-death nature of the physiological data, BSNs are highly dependent on what computer science calls *context awareness*: the ability of a particular sensor or network of sensors to understand something about the environment it is sensing. Context awareness is ultimately then about a computing system's ability to detect a user's internal or external state.[50]

The power to predict is also the secret sauce in the necessary coupling between sensing and AI technologies. But talk of hardware without software and software without hardware is ultimately meaningless, just like discussions of body without mind or mind without body. It leads us to some common fallacies: that *data exhaust*,[51] the trail of discarded data that we leave behind after interacting with a computing system, somehow comes without physical origin, or that sensing is inherently intelligent because it has some basic form of computation embedded into it. In the case of BSNs, context awareness is only possible because physically captured material data from the blood, nerves, or skin is also subject to prediction: fed into the standard and, if not for outsiders, seemingly oracular mathematical techniques, such as neural networks, Bayesian probability, predictive modeling, or hidden Markov models. While these models differ in their mathematical specifics, they all converge around the similar concepts of modeling and predicting behavior statistically.[52]

These different mathematical techniques (and later, when they are turned into instructions for computing machines, algorithms) give (very limited) context to sensors. But context needs not only to be trained but also to be reinforced in sensor networks, given the large possibility of errors generated by sensor noise, mistaken classification, or too many different kinds of sensor data that need to be understood and analyzed. In other words, classifying or discriminating is one of the things our brains are good at doing, but it's not the only thing that helps us distinguish context. Context is also based on "external knowledge sources unavailable to machine systems."[53] There are numerous things that are fiendishly difficult for our sensor networks at present but that human beings by nature quickly grasp to understand context: sensory memory, emotional state, sensitivity to environmental factors, integrating different senses, or even biochronological information that affects repeated moods or sensations, such as the time of day or the season.

Prediction, classification, and persuasion thus go hand in hand when it comes to sensor networks, whether on, near, or below the skin. Most likely, psychophysicist Gustav Fechner and experimental psychologist Wilhelm Wundt could never have envisioned that eventually everyone would carry a miniaturized physiology laboratory around, in and on them.

But to turn our human bodies into continual experimental test subjects that are being monitored 24/7 might just, in fact, be the ultimate apotheosis of the nineteenth-century laboratory, in which the senses could be precisely monitored, qualified, and predicted, all without the intrusion of the experiencing self. As we saw at the start of this book, this is a historical problem that has become so embedded in our technical thinking today that we no longer recognize it as a problem. But when the relationship between sensing and meaning is reduced solely to prediction, when the connection between information and signification is severed, and when distributed sensing mainly denotes continual monitoring, whether from the floor of the ocean or the insides of our organs, we reduce both the scope and potential of both biological and machine systems.

Researchers acknowledge that there is a significant technical difference in the kinds of applications normal sensor networks operate versus the richer specificities and complexities of physiological experience. But it might instead be more useful to give up turning these technologies into direct likenesses of us and start seeing them as other sensing entities that now shape scales of new forms of experience, from the intimacy of the body to the scale of the city. Some sixty-one years later, Nieuwenhuys's vision of the great game to come still provides a lesson: "The investigation of technology and its exploitation for recreational ends on a higher plane is one of the most pressing tasks required to facilitate creation of a unitary urbanism on the scale demanded by the society of the future."[54]

V Enhancing

11 Sensing and Hacking the (Soft) Self

> A self is not something static, tied up in a pretty parcel and handed to the child, finished and complete. A self is always becoming.
> —Madeleine L'Engle, *A Circle of Quiet*

In 1968, a relatively unknown University of Chicago researcher named Joe Kamiya published a four-page article in the magazine *Psychology Today* with the provocative title, "Conscious Control of Brain Waves." Kamiya, who was conducting research in the university's psychology department sleep lab, had long been fascinated with a concept called *self-perception*, the idea that someone has a direct awareness of "features, behavior, and body processes, including their feelings, emotions, thoughts, and memories."[1]

While learning to read the brainwave recordings generated by a cluster of tiny sensing electrodes mounted on the scalp of test subjects, Kamiya noticed the tendency of younger participants in his study to produce irregularly changing brain rhythms called *alpha waves*. Alpha waves are slow, 8–12 Hz oscillations in the brain that usually indicate a state of mental relaxation and that become particularly pronounced when closing one's eyes.

Kamiya raised a question. Was there a correlation between these fluctuating rhythms and the change of a subject's self-perception—that is, their subjective experience? Subsequently, Kamiya commenced a series of experiments to explore whether someone could literally control their brainwaves—an action thought to be completely involuntary at the time—in a voluntary manner directly with one's will. Kamiya hooked up a male graduate student named Richard Bach and other test subjects to a machine that generated an electroencephalogram or *EEG* recording that sensed and monitored the electrical impulses created by neurons firing in the brain by way of dozens of electrodes attached to the subject's skull.

He then tried several experiments to see whether such alpha waves could be controlled through the presence of some kind of feedback mechanism. Kamiya would sound a tone whenever a sudden "burst" of alpha waves took place (which only he knew about) and then asked the subjects to identify whether they were "in alpha" or not. Bach, for one, excelled in an unusual way. Nine times out of ten, he was aware that he was entering into the alpha state.

But it was in a second experiment that something more unusual happened. Each time he heard the tone, Bach was able to voluntarily shift his alpha frequency by almost 1 Hz—a sizable number. This voluntary change seemed to be definitive proof that with the right context, someone could actually influence a physical phenomenon originally thought to be wholly involuntary.

In later verbal interviews, the subjects would describe the experience of being in the alpha state as an overwhelming feeling of "alert calmness," a similar description to that advanced meditators use to describe intense feelings of inner serenity.[2] Yet it was Kamiya's conclusion that voluntary control of brainwaves might also bring one into a new state of consciousness equivalent to that of Eastern meditation traditions that partially opened the floodgates to a new way of sensing the self with machines. What Kamiya had called *neuro-* or *biofeedback* was about to go mainstream.

Research into biofeedback originally arose in the mid-1960s, mainly in clinical contexts. It quickly spread among a disparate group of researchers from California and the East Coast of the US to the UK, where British neuroscientist William Grey Walter was similarly exploring brainwaves in relationship to altered states of consciousness.[3] From a scientific standpoint, a phenomenon like biofeedback was certainly not a given. Studies were vigorously debated, particularly around the question of whether one could actively and voluntarily control alpha waves.

Outside of scientific debate, however, exploring devices that sensed and measured brainwaves soon became a new technological means to not only discover the self, but also actively alter it. After Kamiya's publication, the

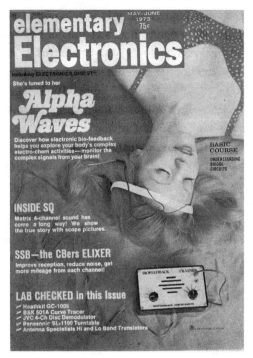

Figure 11.1
Alpha wave trends in the 1970s in the popular DIY magazine *Elementary Electronics*.

interest in using biofeedback to experiment on the self suddenly exploded in the mid-1960s and quickly became a major countercultural trend.

As one chronicler of the time writes, "Word got out that scientists had stumbled onto a wonderful and halcyon state of being that everyone could access, a Shangri-La within the mind."[4] For example, Esalen, the famous Northern California retreat center founded in a ramshackle cliffside hot springs area overlooking the Pacific in 1962, became one of the anchor points in the movement through its workshops concentrated on biofeedback's use for self-realization, rehabilitation, and mind-body medicine.[5]

The technique also attracted the excited attention of experimental artists and musicians like John Cage, Alvin Lucier, and David Rosenboom, who were always on the lookout for new sound sources.[6] As the American composer Richard Teitelbaum wrote in 1976, "The promise of actually 'orchestrating' the physiological rhythms of the human body—heart, breath, skin, muscle, as well as brains—with whatever material from the vast gamut of electronic (and modified concrete) sounds was an exciting one, both musically and psychologically."[7] This interest in electronically inclined musicians and composers to use brainwaves as a new means of music making didn't stop in the 1960s. It carried on from the 1970s to 1990s in the work of a more diverse group of music makers and visual artists, including the American composer Ruth Anderson, who was the first woman to set up an electronic music studio, this one at Hunter College at CUNY in New York. Anderson's 1979 composition *Centering* directly demonstrated the interest in biofeedback as a creative tool that could explore body and self. A group of four audience members outfitted with galvanic skin response sensors observe a dancer. The changing arousal that the audience members experience, which is picked up by the GSR sensors, is then used to affect the rise and fall of electronically generated pitches that constitute the score.[8]

If biofeedback eventually fell out of general favor in psychology and neuroscience in the late 1970s, presumably because of its connection with countercultural ideologies, it was soon to be picked up in a new area of research called *consciousness studies*. These researchers were anxious to prove that, unlike exerting conscious control over one's brainwaves, meditation itself could directly influence the oscillations. This hypothesis lead consciousness researchers to begin establishing links between brainwaves and those excelling in meditation practices, from Indian yogis to Tibetan and Zen Buddhists, with brainwave sensing as the central technological device mediating this relationship.[9]

In 1965, composer Alvin Lucier wore a band studded with electrodes on his forehead and attached to a hulking piece of electronic equipment weighing a ton in order to sense his alpha waves for his live work *Music for Solo Performer*. In 2020, consumer-grade brain wearables or *brain-computer interfaces* (BCIs) look like fashion accessories. They have been released from the lab and, together with fitness trackers, smartphones, and smart watches, now form part of our daily experience in self-sensing. The emerging market for what is termed *wearable neurotech* is flooded with both consumer and professional systems designed to reveal our inner states. Muse, Neurosky's

MindWave Mobile, BrainBit, emWave, OpenBCI, Thync, Emotiv, BioSemi, and mBrainTrain are the names of a few of the devices that eagerly seek to demonstrate that the brain's electrical activity still holds deep secrets to unlocking the mystery of self-hood through sensing and computation. These wearable devices, easily purchased online, aim to create a scientifically valid snapshot of us while providing the power to track and even control our brains.

The interest in shaping the self through the voluntary control of the firings of millions of neurons by way of sensing machines is a long-sought-after imaginary, with each decade seeming to produce a new set of technologies to take advantage of the eternal fascination with the actions of the brain. But the EEG technique that Joe Kamiya utilized in his Chicago laboratory to understand how to consciously shape our mental experience is over 150 years old.

As it entangles sensors and electronics, mathematics and the mysteries of the body, brain and nervous system, the EEG is a prototypical sensing machine. It graphically records the electrical activity of different areas of the brain using a series of surface-mounted sensors, basically electrodes on the scalp. The electrical bursts from single neurons are much too small to be individually picked up by any of these electrodes, which is why these sensors need to measure and sum up the synchronous activity of thousands or even millions of firing or spiking neurons.

The small metal-plated sensors in EEG headsets come in both wet and dry variations. While more accurate, wet electrodes have a distinct disadvantage. For them to have good electrical conductance between the sensor and the skin, they need to be applied with a sticky, sometimes abrasive gel-like substance. In contrast, dry electrodes made of stainless steel, which also conduct electrical signals, are much easier to deal with. They don't require gels and use light mechanical force that enable the electrode to be noninvasively applied to the skin.[10]

The EEG has mainly been used in clinical contexts to detect certain anomalies in the electrical activity of the brain—for instance, what happens at the onset of an epileptic seizure or in the shifting phases when we go into deep sleep. These machines, like the one that Joe Kamiya used, consist of highly fragile components costing upward of tens of thousands of dollars. Because of the high potential of unwanted electrical noise, as well as subjects' movements that can easily interfere with data capture by introducing noise in the signal, clinical EEGs used in hospitals have hundreds of channels for recording brainwave frequencies and sophisticated mathematical processing techniques to handle the avalanche of data produced by thousands of neurons firing synchronously. The electrodes are either preinstalled in caps or, if not, have to be applied in a particular spatial configuration the design of which is actually specified in an international standard—the International 10–20 System of Electrode Placement.[11]

New wearable devices, however, have been driven by the miniaturization of technology; advances in wireless data communication; lower-cost electronics; and more optimized, faster-running software analysis engines. They have also benefited from the use of multichannel dry electrodes installed in headsets, which provide a much more portable and pleasant solution for the wearer.

But emerging electronics, smaller form factors, and portability are only the technical facets of the so-called neurowearables boom. From an experiential perspective, these technologies promise something more profound: new forms of self-actualization that can be achieved by sensing and thus, making visible the inner world of the brain and then acting on these revelations in order to change one's mood or even being.

Canadian firm InteraXon introduced one of the earliest consumer-grade EEG headbands, called Muse, in 2014. Muse affirms that its device opens up

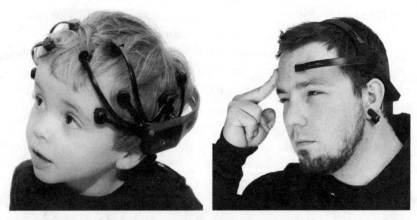

Figure 11.2
The neurowearables boom. *Left:* EmotiV EEG headset. Photo courtesy Thinker Thing. *Right:* Neurosky MindWave. Photo courtesy SparkFun Electronics.

the possibilities for "technology enhanced meditation" to "make the intangible, tangible." Wearing a stylish Muse headset consisting of four dry electrodes embedded in a futuristic plastic shell will enable a transformation of consciousness by giving real time feedback on your brain and body activity to let you know when you enter a certain zone of consciousness. Not to be outdone, Muse not only features EEG possibilities but is also loaded with other sensors, including a plethysmograph, accelerometers, and gyroscopes to capture heart rate, breathing, and motion and balance—or, as Muse puts it, "More Calm, Sharper Focus, Better Sleep."[12]

Muse is not the only consumer EEG device in town. Neurosky, a manufacturer of both professional-grade EEG systems and a consumer headset called MindWave Mobile, uses the branding tag line, "Body and Mind. Quantified." Macrotellect's BrainLink Pro, meanwhile, allows you to "Know Your Thoughts. Be More." BrainBit announces that you will be able to "Use your brain as never before."[13]

The concept of consciously "using" the brain to alter behavior, improve sleep, or bring about a meditative peace of mind is powerful and seductive. Consumer-grade EEG headbands seek to mitigate the killer factors of everyday life by reducing daily stress, improving sleep, and stimulating the brain. Muse, in fact, has an even larger goal in mind: to make meditation easy by providing its users with real-time feedback about their brainwaves, thus forging a stronger mind-body connection. The market and interest for brain-computer interfaces is in fact increasing, indicating that the desire to control the brain seems to tap into something deeper in our desire to shape the mind.

At first glance, the claims of manufacturers that you can "take control" of your brain seem little more than standard marketing speak: wearable technology start-ups seeking to capitalize on putting the newly achieved powers of sensor miniaturization onto a consumer's body in order to unlock the ability to control and modulate one's inner self. If this is so, why is it that EEG, a technology long used in only medical research labs until very recently, is so interesting to us now? How does it fit into our current world of sensing machines?

The technology behind the EEG has a long historical trajectory. In 1875, the forgotten English physiologist Richard Caton is claimed to have done the first neurophysiological studies on monkeys, dogs, and rabbits using a *galvanometer*, an early electrical device that measured changes in electrical current. Caton's work would then have to wait another fifty-four years before it resurfaced, this time in the research of a German psychiatrist who sought to unlock the hidden secrets inside human brains.

Hans Berger was interested in brain phenomena extending beyond hardcore psychophysics and into the more mystical realms of the mind, which included telepathy and what he called *psychic energy*—a measure of the relationship between energy transformation caused by blood flow and brain metabolism, together with mental feelings and emotions. Berger thus established that such psychic energy was expressed through the different electrical frequencies or waveforms that the brain was producing, something he obtained by the use of a larger string galvanometer, what he called a *brain mirror*, which worked as a device to literally reflect the mental actions of the brain.[14] Later labeling these brainwaves *alpha* and *beta*, Berger argued that such electrical signals corresponded to changes in mental state. In fact, alpha waves were later named *Berger waves* by famous British physiologist Lord Adrian, who brought Berger to worldwide fame after Adrian's own EEG tests in Cambridge in the 1930s obtained similar results.[15]

The five core brainwaves of gamma, beta, alpha, theta, and delta indicate different frequencies of neural oscillations that very roughly correspond to different states of brain activity. Ranging between 32 and 100 Hz, for example, gamma waves appear during periods of heightened concentration, such as problem solving. Slower delta waves oscillate between 0.5 and 4 Hz, and occur mainly during sleep, indicating a loss of awareness. In fact, when we fall asleep, we begin to cascade down through all of the brainwaves, moving from beta through alpha and theta until we finally reach delta.

Although there are decades of research into these oscillations, neuroscientists are still not exactly sure how this electrical activity really corresponds to different mental states of brain activity. Despite this gap in knowledge, however, the slower alpha waves would quickly become a kind of holy grail in the popularization of brain science, mainly because they constitute a dominant oscillation that indicate states of mental relaxation. Joe Kamiya's 1960s experiments into biofeedback, for example, sought to understand how these different waves might be influenced by way of self-directed focus. His work resurrected Berger's more esoteric, and yet entirely seductive idea that some form of psychic energy might be found in the quantification of the mysteries of brain activity.

This very abridged history demonstrates that the practice of "taking control" of one's brain has long existed on the fine line between legitimate neuroscience and esotericism. While neuro- or biofeedback has been utilized in clinical contexts for treating everything from sleep deprivation and depression to attention deficit disorder, the claims of technological truth-telling achieved by tapping into one's inner mental states to achieve bliss through direct-to-consumer EEG headsets are more contentious. Studies that challenge the idea that the sensors that make up a consumer neurowearable will deliver anything remotely resembling a clinically accurate reading of brainwaves increasingly appear. One report found that effectiveness claims about portable EEGs, ranging from the somehow realistic-sounding ability to enhance memory and cognitive skill to more outlandish ones announcing weight loss reductions by eliminating cravings for food are rarely supported by scientific evidence, let alone larger studies.

An even more interesting warning appeared in another study. If neurowearables claim to improve sleep and heighten cognitive skills, what is to prevent consumers from continuously wearing such devices to continually self-improve 24/7? At first this sounds harmless. Consumer EEG headsets, easily bought online, are essentially noninvasive technologies. But another neural interface technology that is increasingly imagined to be grouped with EEG—so-called transcranial magnetic stimulation (TMS)—is anything but.

TMS systems don't sense brainwaves. Instead, they artificially stimulate and modulate the brain's neurons by delivering magnetic pulses to the head underneath the worn device. The technology works on a principal called *electromagnetic induction*, discovered by English physicist Michael Faraday in the mid-eighteenth century. One research report even argued that TMS is a kind of "Faradization of the mind."[16]

Faraday's induction principle is complex mathematically but simple in its conception. It states that a moving magnetic field can generate an electrical current in a nearby material. TMS uses a magnetic coil that is placed against a subject's head. A single or sometimes repetitive burst of electrical current is then passed through the coil, which then produces a localized magnetic field that is transferred from the device into the skin and skull. The resulting effect is a newly introduced current in the brain's electrical operations.[17]

To put this another way, TMS "massages" neurons in the brain. But there is also increasing interest in sensing the results of these newly introduced electromagnetic stimulations of neurons by combining EEG with TMS electronics in the same device. As TMS cranks up the stimulation intensity, the brainwave sensors stand by to measure the change in neural response.[18] TMS is still mainly used in medical contexts, but combining sensing and stimulation in a portable consumer headset is on the horizon. Already you can troll the internet for DIY schematics to build your own TMS system.

TMS forms a kind of essential toolkit for a new cultural movement dubbed *brain hacking*.[19] Brain hacking is essentially like building your own DIY sensing machine—using sensors and actuators installed in brain-computer interfaces, with the goal to shift cognitive functions through a short-lasting but intense burst of energy and then visualize and analyze the results using EEG.

Like cookbooks, dozens of publications, blogs, websites, and YouTube videos have appeared that provide recipes on how to harness the power of the brain. The internet overflows with testimonials of brain hackers seeking to combine acts of self-experimentation with self-observation, and websites abound on how to build one's own DIY EEG or TMS circuit or to hack existing technologies. There are also more serious endeavors in the form of software and hardware technologies that are readily available for brain hacking—for example, OpenBCI, an open-source software library that allows amateur hackers or "anyone interested in biosensing and neurofeedback to purchase high quality equipment at affordable prices."[20] Supported by a range of university and corporate research partners, OpenBCI users can build and 3D-print non-medical-grade EEG headsets and utilize neurofeedback software tools for deciphering and visualizing the complex electrical waves in the brain. With such widely available hacking tools, one can see one's brain activity in real time while feeling stressed or meditating and be encouraged to adapt what one is doing to "get better results."[21]

Brain hacking is not just for individual users. Some hackers are also interested in exploring brain-to-brain communication, in which one set of brainwaves is used to alter the action of another set. Through a brain-to-brain interface, electrical signals from one brain are stimulated and sent via a computer-to-brain interface to another brain to enable a new kind of neural controlled, bidirectional communication.[22]

Strategies for hacking the brain do not only involve TMS to potentially improve problem-solving abilities, focus concentration, or tune the brain better. Apps readily available on a smartphone that can capture data and sync with wireless neurowearables do what experimental biofeedback composers struggled to do in the 1960s: they translate or "sonify" brainwaves into music, bird sounds, or other immersive soundscapes, like the weather. The Muse headset maker announces, "Wearers will hear thunderstorms when their mind races, and the sounds of soft crashing waves or birds if they are calm. The idea is to try to calm the storm, with your mind, to reach a more-relaxed state."[23] The inevitable gamification of these apps also enters the picture, with users working to reach goals and milestones and being rewarded with points and scores. That is, mental states like hard concentration, sitting calmly, or daydreaming are recast as playable games with rewards.

⌒

By now, you might be asking yourself: How can such hacking with readily purchased EEG technology or build-it-yourself TMS systems actually enhance our cognitive activity? In clinical contexts, TMS shows promise in treating mental health issues like depression and attention deficit disorder, as well as being an important tool for studying brain development. But the key grail that hackers really seek to transform is a characteristic of brain function called *neuroplasticity*.

Neuroplasticity is a deeply fought-over term. *Plastic* signifies that the brain is capable of change. Such change, however, comes in two varieties: *functional*, which involves the transformation of nerve cell activity or function, and *structural*, in which new connections between neurons (synapses) and new neural pathways between different groups of neurons are generated together with changes that take place across entire brain regions. The time scales for neuroplasticity vary as well, from milliseconds to hours, days, or even decades.[24]

Brain hacking's ultimate goal is to use TMS or other methods to affect such neuroplasticity—in effect, to rewire the brain to alter it and, hopefully, improve general behavior. But this is not the only way that neuroplastic transformations can occur. Besides TMS, one of the most interesting and controversial ways of manifesting neuroplasticity lies in creating technological devices called *sensory substitution* systems that can help jump start the neuroplasticity process or, at the very least, aid in its development.

Sensory substitution is a much-debated concept that describes how the sensory system can translate an existing sensory neural pathway or substitute it for a missing or damaged one. While sensory substitution can occur naturally in the brain, researchers have long been creating and building machines that aim to facilitate such translation.[25] As one article stated, "'Sensory substitution' denotes the ability of the central nervous system to integrate devices of this sort, and to constitute through learning a new 'mode' of perception."[26]

American Mexican neurologist Paul Bach-y-Rita is credited not only with discovering this new mode of neuroplastic perception but also with developing the first kind of sensing machine for it to take place. Bach-y-Rita's earliest apparatus, called the Tactile Vision Substitution System (TVSS), dates back to research he and collaborators conducted in the mid-1960s at the Smith-Kettlewell Eye Research Institute in San Francisco. This research explored how artificial sensors might be used to develop a "visual substitution system for the blind."[27] The system would send visual information by way of an *artificial receptor*—an array of sensors—to the brain.[28]

The artificial receptor in this case was a bulky and early black-and-white video camera that could translate and project what the researchers called *mechanical television images* directly onto the backs of blind participants. The original apparatus now looks antiquated and bizarre—a mix between steampunk and the aesthetics of the 1987 film *Brazil*.[29] Consisting of a used dental chair, the camera was attached to a tripod and hooked up to an even more bulky device called a *commutator*, a machine that translated the electronically scanned images from the camera into individual pixels (figure 11.3, left).

These digitized images were then fed to a twenty-by-twenty array of four hundred small, metal-sheathed actuators or solenoids that were embedded into the back of the chair. Operating at around sixty vibrations per second in proportion to the brightness of the pixels in order to move back and forth, the actuators converted the camera's image into direct vibrotactile stimulation delivered onto the backs of the experiment's blind participants.[30]

Bach-y-Rita's sensing machine proved remarkable. Within an astonishingly short time, a period of minutes, the blind subjects were able to identify

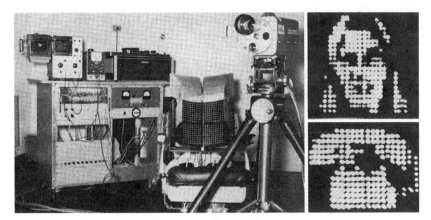

Figure 11.3
Tactile Vision Substitution System from Paul Bach-y-Rita and associates, 1969.

how many points were simultaneously being activated on their skin. After an additional five minutes, they could tell the difference between primitive geometric shapes. Another test, this time aimed to describe the difference between the sweeping movements of a horizontal versus vertical bar, quickly yielded 100 percent identification.[31]

Subsequently, the subjects were "shown" a "vocabulary" of twenty-five different objects, from coffee cups to stuffed animals, generated by different vibration patterns from the actuators and asked to "scan" and identify them using the camera (figure 11.3, right). As the number of repeated showings of the same object increased, the amount of vibration on the back needed to identify that object decreased. The objects were then placed on a table in front of the subjects who were asked to manipulate the television camera in order to properly distinguish between them. Here, the complexity of the real world took over. Characteristics of vision that most of us take for granted became glaringly apparent: occlusion (blocking) of objects; shapes that became distorted or changed their size as one shifted perspective.[32]

What this simplified account suggests is that the subjects perceived the objects because they felt their shape, form, and contour as patterns on the skin. But the experiential descriptions given to the researchers from the subjects depicted something entirely different. After sufficient training and learning, the subjects began to describe the objects as being placed in space—literally projected in front of them as "something out there" rather than the feeling of distinct patterns on the skin.

Bach-y-Rita argued that a machine like the TVSS worked precisely because of the brain's plasticity—that the brain could reorganize its functions despite damage or deformity. This reorganization would stem from the complex nature of how the brain receives and decodes sensory information. The brain almost worked like a traffic detour, rerouting processes around a region that may no longer be able to function.

Throughout his lifetime, Bach-y-Rita continued to iterate his basic machine with different features until, in the late 1990s, he and his graduate students at the University of Wisconsin developed a commercially wearable system called the BrainPort. The device already went far beyond today's wearable brain-computer interfaces, allowing video images produced by a camera worn by a blind user to be directly translated as electrical stimulations on a new and unusually sensitive place—namely, the user's tongue.[33]

While these descriptions of tongue-based imaging and brain computer machines that reroute neural signals from damaged parts of the brain sound like the stuff of science fiction, sensory substitution has increasingly gained traction in the popular imagination. A 2017 *New Yorker* magazine article brought Bach-y-Rita's early work, culminating with the BrainPort, to the attention of a broader audience.[34] Moreover, while the BrainPort is now an almost fifteen-year-old commercially available device, newer machines such as EyeMusic, EyeCane, vOICe, VEST and others that translate text into sound or camera-based images into vibrations across the skin are appearing in the growing markets for sensory substitution machines.[35]

Given brain hackers' interest in altering neuroplasticity with machines, we should have expected that sensory substitution would not remain strictly within medical or assistive technology contexts but would inevitably also attract military interests as well. *Emerging Cognitive Neuroscience and Related Technologies*, a 2008 US National Research Council report focused on United States security issues, discusses sensory substitution and plasticity, together with detailed overviews of neuropsychopharmacology, cognitive biology, and human-machine interface research. The report featured a historical overview of devices that "improve or extend human performance in the cognitive domain through sensory substitution and enhancement capabilities" but also suggests new mergers between sensory substitution devices and brain-computer interfaces to enable the "controlling of weapons by sensory and cognitive means."[36] If stories circulate in social media about brain wearables enabling you to control everything from cars and drones to

TVs and games with your mind, evoking the subtler edges of telekinesis and other parapsychological phenomena, a more conspiracy-driven, dark web version of these sensing machines lurks in the background. In this sense, the counterculture of brain hacking is as fascinating as it is downright scary.

On one side, researchers claim that in the next twenty years, "neural interfaces are likely to be an established option to enable people to walk after paralysis and tackle treatment-resistant depression, they may even have made treating Alzheimer's disease a reality."[37] Yet for every website that announces the power to control your brain for enhanced mental function or treating vicious illnesses, another claims that forces are out there waiting to steal your thoughts. Sites at the fringes of the internet announce that hackers or, even worse, advertisers could in the very near future implant messages into your thoughts. By using "spyware on your mind," brain hackers could thus gather private data, such as political preferences, sexuality, or PINs.[38]

Even brain hackers' goal to rewire the brain might produce unknown consequences with regards to neuroplasticity.[39] Neuroscientists are still unsure whether zapping different brain regions with electromagnetic fields, as one can do with TMS to influence cognitive function, might not lead to mutant neuroplastic transformations—to a kind of out of control change in neural structures.

EEG headsets are not the only consumer sensing technologies to have walked out of the laboratory in the past years to exploit our growing interest in altering the self through sensing machines. Now we also have to contend with wearables that "invade the dreamworld"—new technology that uses sensors attached to our bodies to record our movements while we sleep and that triggers light, sound, vibration, and other stimuli to bring on or influence how we sleep and dream.[40]

Sleep and dreams are the ultimate frontier for sensing machines. Sleep is where we spend almost twenty-five years of our lives. It is not only a potential $32 billion dollar market. In our stressed-out, increasingly insomniac existence, sleep is also increasingly seen as the latest status symbol.[41]

It certainly makes sense that the sleep sensing industry is booming. According to the WHO, the lack of sleep in the most advanced industrial countries is essentially seen as a viable health epidemic, catalyzing diseases such as cancer, obesity, and heart problems while also contributing to mental illness.[42] The economic loss is even more devastating, with a RAND study of advanced industrial nations arguing that the impact of lost productivity on the US economy alone is around $411 billion per year.[43]

With these almost extreme conditions, it's no wonder that sleep is forecast to be the next big thing. According to one estimate, the market for sleep should reach over $100 billion dollars by the mid-2020s.[44] If this enormous number is any indicator, our sleeping and dreaming are poised to be transformed by all manner of sensing machines that will accompany us into the mysterious universe of slumber.

The phenomenon of trying to monitor and influence sleeping and dreaming will sound strangely familiar. Just like portable EEGs or other wellness devices like wristbands and mood rings that can measure your emotions based on how you sweat, worn and app-based sleep trackers also promise new possibilities for self-optimization. These gadgets sport branded names that sound as if they were generated by automated algorithms: S+, Beddit, Basis, Oura Ring, Emfit, Sleep Cycle, and EarlySense. Like brain-computer interfaces, sleep trackers appeal to scientific evidence for validation. They aim to bring a new sense of calm, reflection, relaxation, and self-realization. What's more, smartphones are even making separate hardware obsolete: you don't even have to buy another wearable gadget but simply can download one of the many available apps that harness your phone's accelerometer to monitor everything from sleep quality, duration, and snoring to sleep phases and environmental factors, like light or air quality.

Sleep trackers also announce grandiose claims about their effectiveness. One product, EverSleep 2, even states that it is "like having a sleep lab in your own bed," a rather exaggerated statement considering real sleep labs have medical-grade EEGs and other monitoring systems, together with rigid protocols for gathering and storing personal data for analysis.[45] In fact, one reason large EEG machines occupy sleep labs is that brainwaves are the most accurate data to tell us when we enter the different sleep phases and what might roughly go on within them. In this sense, the sleep trackers you can buy online or at Walmart function radically differently from even consumer-grade wearable EEGs, let alone medical-grade systems. They don't measure sleep but rather how much you move during sleep. That is, these trackers assume a basic formula: the more you move, the less likely you are to be slumbering.

How can your phone determine how well you might or might not sleep? In clinical contexts, sleep monitoring has long been based on a complex, time-consuming process known as *polysomnography*. The basic aim in polysomnography is to use multiple physiological sensors, beginning with an EEG and complementing that with breathing, heart rate, electrical activity in the muscles, and a range of other data in order to detect and diagnose sleep disorders, from sleep apnea (sleep-related breathing disorders) to twitching limbs and insomnia.

Given the multistage nature of sleep and the mathematically complex analysis and correlation of multiple streams of sensor data, it would thus seem ludicrous that a $150 wrist band could actually detect the differences between different sleep stages. Yet hundreds of apps have been developed to take advantage of accelerometer-measured movement data that happens during sleep by utilizing a more efficient and less costly process called *actigraphy* in order to alert the user about their overall sleep quality.

Actigraphy studies a foundational part of the sleep process—the ongoing cycle between rest and activity—by utilizing motion data from the accelerometer to indicate features such as total sleep time, percent of time spent asleep, total waking time, percent of time awake, and other characteristics. The accelerometer records motion over specified time windows that correspond to the presence or absence of movement. In post analysis, these indicators are then used to deduce what sleep cycle a person may have been in. Of course, the distinguishing feature of each device is the propriety of each manufacturer's algorithms. A filter that is tweaked better or an equation that extracts features a bit more efficiently might set one device slightly apart from its competitor. It now seems run-of-the-mill that every major fitness tracker corporation is vying for the best engineers, medical doctors, and mathematicians so that their algorithms stand out above their neighbors.[46]

Amazingly, even though the manufacturers of sleep trackers claim to understand sleep through actigraphic studies, neuroscience itself still does not know what the exact purpose of sleep is. According to British neuroscientist Matthew Walker, whose 2018 popular science book *Why We Sleep* gives a devastating account of our current lack of it, sleep is pluralistic. It covers a multitude of ground, restoring fundamental brain and body functions, from memory to cardiovascular health. Some researchers argue that sleep is the antidote for too much synaptic activity. In other words, sleep is about forgetting, reducing the noise spawned and swarming in our heads due to the brain's machinations so that we can heal the synaptic overload that occurs during waking hours.

When we learn, new synaptic connections are forged and strengthened between thousands of neurons. But this synaptic strengthening has its costs, including stress on the cellular systems and changes to even less understood support cells, like the *glia*—the mysterious, jelly-like nonneural substance that seems to hold and protect the neurons in our brain. We therefore sleep not only to disconnect from the waking world. We also sleep because we need to reduce the amount of neuroplasticity, the constant adaptations of the brain, that the organ is subject to in daily life. Sleep, in other words, is "the price we pay for neuroplasticity."[47]

Like other sensor-driven industries, a myriad of sensing machines exist that can track, lull, or monitor us in sleep. These include both wearables and *nearables*, devices that conveniently occupy the night table. Alas, before you pick up this book, however, most of these start-up-driven devices will have ceased to exist or will have changed their consumer orientation altogether. Take Sense, for example. A glowing, polycarbonate sphere resembling a bird's nest, with the form factor of a tennis ball, this device loaded with environmental sensors measuring light, sound, humidity, and temperature and communicating with a 6DoF accelerometer device literally called the Pill, is already no more. Created in 2016 by an upstart start-up called Hello run by a twenty-two-year-old inventor and briefly valued at over $500 million USD, the company with its all too futuristic sleep machine that would change color to indicate what quality of sleep one was in went bust after failing to be acquired by Fitbit.[48]

Another wearable called Dreem 2 still exists at the time of this writing. An EEG headband, Dreem 2 pitches its innovation based on the number of sensors it has: four dry electrode sensors to measure brainwaves, a pulse oximeter to gauge pulse, and an accelerometer, in addition to bone-conducting haptics that causes your bones to resonate through mechanical vibration. Interestingly, however, Dreem's branding and focus radically changed since its debut in 2016. Its earlier website revealed a slick-looking device designed by Yves Béhar, a well-known product designer and featured high-definition videos of flowing fabrics and undulating forms. In 2021,

Sensing and Hacking the (Soft) Self 237

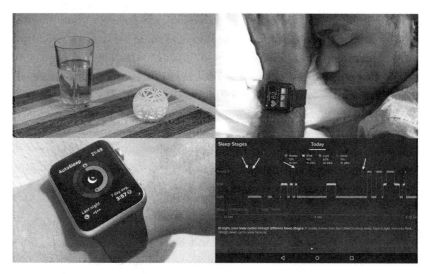

Figure 11.4
The sleep sensor and tracking industry. *Top left:* Sense sleep-tracker. Photo by original daniel. *Top right:* Sleep tracking. *Bottom left:* Sleep tracking on an apple watch. Photo by integratedchange. *Bottom right:* Fitbit sleep dashboard.

the site was rebranded with the tagline "Building the healthcare of tomorrow" and focused on the use of the device for clinical trials with lists of academic partners worldwide.

Other devices deploy vibration, sound, and light to stimulate our senses in order to encourage sleep. THIM, a bulky ring worn on the finger and coupled to an app, is advertised as "from the sleep lab to your home" and administers vibration based on scientific sleep trial techniques in a procedure that wakes you up after you fall asleep several times in order to encourage you to sleep better. Re-Timer, another device from THIM's inventor, is a similarly worn product resembling a pair of augmented safety glasses. It uses pulsing green-blue LEDs to artificially reset your circadian rhythms and supposedly to suppress melatonin production so that we can change the time we feel tired. Re-Timer appears to be based on the concept of *light therapy*—the idea that in light-deprived climates, one can stimulate the circadian cycles artificially to compensate for the lack of white light in the natural environment.[49] As with so many of these devices, there is a host of competitors. PEGASI, AYO, FeelBrightLight, and others, many jump-started by Kickstarter campaigns or seed funding, all purport to transform your sleep using colorful, LED-based technologies.

There is something ironic about the sleep-tracking industry. It is increasingly reported that we cannot sleep due to the continual intrusion of media and electronically laden devices into our lives, and yet the sleep market is saturated with these same devices aiming to help us sleep.[50] In fact, studies are emerging that argue that our very obsession with sleeping better is leading to a new sleep disorder called *orthosomnia*—literally, a kind of reinforced sleep-related anxiety caused by an unhealthy obsession with achieving the perfect sleep that increasingly becomes impossible to achieve because of these devices.[51]

Although they involve the latest in dry electrode sensing and machine learning to feature-extract, classify, and recognize sleep patterns, sleep trackers are the soberer form of a much weirder subgenre of devices: so-called brain machines. Brain machines are devices designed to stimulate or entrain the brain by generating patterns of light or sound that encourage neurons to oscillate or synchronize at the same phase and frequency as such audiovisual patterns. A quick search online reveals all kinds of wonderful machines that seem to occupy the space between UFOs, Tantric Yoga, and the tinfoil hats designed to cancel out radio wave thought control. Mind-Machines.com claims to "specialize in providing theraputic [sic] tools for relaxation, enhanced learning, mind power, biofeedback, neurofeedback, high tech meditation and personal achievement." Clicking on the "Brainwaves" button, one finds more exotic taglines, such as "Imagine Automatic Mind-Yoga at the touch of a button! With The Brain Entrainment *Light Sound Mind Machine* with binaural beats . . . it is now possible!" In addition to entrainment, these brain machines also are claimed to "expand your natural alpha brainwaves and theta brainwaves using light and sound." Whereas "beta brainwave stimulation" seems to heighten "athletic performance," delta waves are useful for "deep, restive sleep."[52]

Another scan of the internet reveals dozens of these sensing sleep machines that mix measurement and stimulation within the same device. The Luuna Brainwave Brain Sensing Bluetooth Smart Sleep Mask, for example, is an EEG-based headband/sleep mask that turns bio-data into soothing music to allow you to sleep faster. Hupnos is a "self-learning" sleep mask that alerts you when you snore, while REMzen monitors "ocular muscle activations"—muscle fluctuations in the eye—to wake you up when you're in a light sleep state.

These devices most certainly live in the netherworlds of established science. But is there really a large difference between them and commercial EEG headbands that promise similar things? Whereas in the past flickering glasses and glowing orbs would have been associated with new age ideology, standard consumer brain wearables that promise new views into our selves now explicitly attempt to legitimatize themselves via direct appeals to research and science. From smart pillows playing ambient sound and smart therapy lamps pulsing away based on what sleep stage you enter, we thus seem to employ all manner of artificial media to assist us in not only judging whether or not we are sleeping correctly but also actively putting us into a state in which we can actually sleep.

Joe Kamiya's original forays into biofeedback research took place in the sleep lab of the University of Chicago's psychology department. This makes sense as sleep labs are locations to sense not only sleep but also something far much more enigmatic: namely, our dreams. Ever mysterious and unique to each individual, the dream, wrote psychologist and dream champion Carl Jung, "shows the inner truth and reality of the patient as it really is: not as I conjecture it to be, and not as he would like it to be, but as it is."[53]

Dreams are essentially stories, impressions and sensations the brain tells us while we slumber, generated mostly during REM, the fourth and deepest phase of sleep. Although it has been demonstrated that dreaming occurs in non-REM sleep as well, in REM the brain is most active, and our pupils nervously flutter back and forth behind closed eye lids. The secret of why and what we dream, however, still remains a fundamental and deeply intriguing mystery, locked up in our physiology, history, genes, and culture. Endless theories abound about the *why* of dreams: The brain tells stories to organize memory and impressions that it picks up during the day. The brain pulls up the impressions and desires that are latent but buried in the unconscious. Or the brain simply generates mere epiphenomena as a by-product of its electrical and chemical processes.

Like sleep, researchers also monitor dreams using EEG recordings of brainwaves, as well as with other physiological sensors—such as electromyograms

(EMGs), which measure small electrical discharges when you move your muscles. Because your muscles are essentially paralyzed during REM in a phenomenon called *muscle atonia*, EMG sensing is used to try and detect low muscle movement which should occur at the onset of dreams. But in the 1980s, a Stanford University–based psychologist named Stephen LaBerge with an interest in dreams took the next step.[54]

Le Berge developed an unusual device consisting of a pair of goggles installed with an infrared emitter-detector pair that could monitor rapid eye movement occurring during the REM sleep state. The device, named DreamLight, was able to sense when subjects would enter REM and, consequently, flash tiny amounts of light into their closed eyes by way of small incandescent lamps mounted in the glasses. In interviews after such experiments, subjects revealed something quite astounding. Not only did they report an awareness of their dreaming (a so-called lucid dream) but they also saw certain lighting flashes and patterns that were generated by the DreamLight directly inside their dreams. In other words, LaBerge was not only interested in measuring dreams. He wanted to influence them using a crude worn sensing machine to create feedback and interaction between the device, the self wearing the device, and the inner experience of that self.

While LaBerge's experiments took place in the 1980s, the devices that now fill Kickstarter campaigns and Alibaba shopping carts from lucid dreaming "instigators" are strikingly similar to LaBerge's, with one exception: they use better and cheaper sensors that allow you to "take better control of your dreams." These lucid dreamer devices use EEG to monitor when we are in REM and then trigger light, sound, vibration, or even tiny, mild transcranial electrical stimuli to instigate lucidity. More recent forays into "dream engineering" in research labs, like the Fluid Interfaces group's Dream Lab at the MIT Media Lab are basically trying to achieve results similar to LaBerge's. Emboldened with the title Engineering Dreams, a particular Dream Lab project describes building "technology that interfaces with the sleeping mind. As the dreamer descends into sleep, we track different sleep-stages using brain activity, muscle tension, heart rate, and movement data. External stimuli in the form of scent, audio, and muscle stimulation affect the content of the dreams."[55]

But tracking dreams is not the Dream Lab's only goal. The authors of a technical paper about the group's research platform, appropriately named Dormio (the Latin derivation of *dormire*—to sleep), write that "sleep is a forgotten country of the mind.... Sleep offers an opportunity for prompting creative thought in the absence of directed attention, if only dreams can be controlled."[56] Yet how does one actually control one's dreams? The word *control* is a misnomer here considering that the MIT researchers use techniques very similar to those that LaBerge's dream device deployed, including sensors to indicate when one falls asleep or audio phrases to prompt the dreamer while they are dreaming.[57]

But the Dormio project also brings to mind another contemporary example of dream control: Christopher Nolan's 2010 film *Inception*, about a professional thief or "extractor" who implants information into the subconscious of others. The MIT graduate students who developed the research indeed claim a piece of the inception heritage when they write "with this new Dormio system in place we are able to influence, extract information from, and extend hypnagogic dreams."[58]

For the moment, the Dream Lab and other dream research projects focus mainly on academic research utilizing standard sensing/actuation approaches to monitor and then to try and exert or "plant" some kind of external media influence into early sleep stage dreams.[59] The endless discussions online and in the scientific literature reveal that *recording dreams* mainly means recording the physiological side effects that are produced in the body while dreaming: nerve impulses flowing through the muscles, eye twitching, bodily fidgeting, decreased blood flow. In other words, sensors register the outcomes of a dream but not the dream itself. The sensors that adorn these dream-based interfaces cannot record what is going on inside your head. The science fiction visions of dream technology infiltrating films from the 1990s, like Douglas Trumbull's *Brainstorm*[60] and Katherine Bigelow's *Strange Days*,[61] in which head-mounted brain-computer interfaces called "SQUIDs" record dreams and play them back in an altered state, seem very far off indeed.

There is no doubt that an uncanny feeling arises the longer one browses through websites or research papers detailing wearable EEGs and lucid dream machines that stand ready to be of service in industries that go beyond wellness and self-actualization. In fact, external sensors that measure brainwaves, eye movement, electrical changes in the beat of the heart, the tensing of muscles, conductance on the skin, and even facial recognition now are used so frequently in *neuromarketing*, the study of the brain to predict and potentially even manipulate consumer behavior and decision-making, that these applications are seemingly yesterday's news.[62] Newer projects on the event horizon suggest that getting into your dreams directly with sensor technology might be just a matter of time. Implantable systems that operate directly within the central nervous system—so-called next-generation neurotechnological cortical-based brain-computer interfaces—may soon further blur the line between artificial and biological electrical signals, making them both almost interchangeable.[63]

Technology utopians like Elon Musk are reportedly working on implantable sensor-driven brain-computing systems. With job postings from 2019 for surgical technicians, animal care specialists, and roboticists, Musk's much ballyhooed company Neuralink seems at the moment to be the most out there in terms of vision.[64] Neuralink aims to create no less than a highly scalable interface composed of ultrafine polymer electrodes or threads, a neurosurgical sewing machine–styled robot to implant these threads, and custom high-density electronics that have massive carrying channel capacity that can be implanted in direct proximity to real neurons. Neuralink's vision is that of an implantable surrogate brain that gathers data via thousands of microsensors directly from its physical partner.

Neuralink may have serious competition in the near future in the guise of DARPA's Next-Generation Nonsurgical Neurotechnology research program. DARPA is well known for its interest in cognitive research, but its investment in new wearable BCIs seems to take the next step: "a more accessible brain-machine interface that doesn't require surgery to use, so that DARPA could deliver tools that allow mission commanders to remain meaningfully involved in dynamic operations that unfold at rapid speed."[65]

The list of DARPA-funded research themes and projects almost boggles the mind: nanotransducers that can be linked to neurons; light and acoustic sensors that enable read/write operations at high speed/resolution directly at the neuronal level; a project from the applied physics laboratory at John Hopkins University that proposes an "optical system for recording from the brain," where path-length modulations in neural tissue could be directly correlated with neural activity.

Another research team at Xerox PARC is developing neuromodulation devices that couple ultrasound waves with magnetic fields to generate localized electrical currents in order to "write" data directly to the brain.

A team at Rice University in Houston, Texas, goes even further, purporting to develop technology that records from and writes to the brain by trying to infer neural activity through measuring the scattering of light in neural tissue and to make neurons sensitive to magnetic fields in order to write to them.

One becomes frightened in just imagining the increased surveillance and threats to individual privacy suggested by such systems. Indeed, there is no telling where these neurotechnologies are headed, but one thing is clear: they promise the ultimate dream of interfacing directly into the grey matter of the brain and, hence, into life itself.

Between EEG sensing and neuroplasticity, observing our alpha waves, and hacking into brains and dreams, it seems challenging to fully grasp the extent to which sensing machines alter the conscious and unconscious experiences that we believe constitutes the self. Some social scientists suggest that despite increasingly miniaturized and complexified electronics and mathematical models, the very belief that brainwave sensing and observation might give us a peek into ourselves replays our oldest dualistic battle between the mind versus body. Feelings and emotions that are not only the product of the brain get translated into thoroughly "objective" scientific concepts, such as neurons and frequency oscillations, which we somehow believe will reveal meaning in and of themselves without regard to context or cultural difference.[66]

Through numerous interviews with alpha wave trackers, one scholar, a Dutch sociologist named Jonna Brenninkmeijer, tried to understand firsthand the obsession with sensing brainwaves. Her interviews and subsequent study revealed that people seek neurofeedback for multiple and contradictory reasons. Some believe that wearable EEGs will help transform them with minimal effort—a kind of instantaneous enlightenment.[67] Others desire what is known as the *placebo effect*: by experimenting with EEGs, they see a path to self-transformation because this is what they have been promised in the advertising of manufacturers and testimonials of users.

Like all sensor-generated signals, however, brain electrical activity picked up by EEG machines is dependent on mathematical techniques, significant signal-processing black magic, which has little connection to such emotionally loaded experiences as trauma or ecstatic feelings. These techniques not only involve filtering, which removes noise generated by the brain itself and by electrical fields near the sensors, but also employs postprocessing techniques like neural networks that attempt to classify or group the brain's mysterious signals. As users of cognitive technologies like EEG headsets or brain machines, we are thus both the watchers and the watched. We observe only the mediated output of our brainwaves, output that can be continually shaped by interaction between us and the electronic hardware and mathematical models that process the tiny mass of electrical signals being constantly fired. The quality of the electronics, the signal-processing technologies, and the context they take place in will determine exactly how and what truthful reading of the brain we can obtain. If we have noise in the circuit, does that mean that we have that same noise in the self?

Even the delta, alpha, beta, gamma, and theta waves that EEG sensors can record are not the whole story. These different waves may be so influenced by similar emotional states that it is difficult to perceive which frequency band the experience of a feeling like alert calmness or other reported experiential states actually takes place in. Is it only my alpha waves that produce and indicate this experience? Is it a combination of waves? Or is it the noise in the device combined with my interaction and with the system that generates the output?

What's more, the very notion that brainwaves made visible by EEG sensors can singularly represent cognitive processes and, hence, self-hood itself is vehemently challenged. A growing chorus of philosophers, cognitive scientists, linguists, and even some neuroscientists argue that our understanding of self-hood cannot be reduced solely to brainwaves or to the heartbeats, muscle tension, sweat, or breathing patterns that physiological sensors capture.

"Outsourcing the Mind," "You Think with the World, Not Just Your Brain," *You Are Not Your Brain*, and "meaning just ain't in the head": These audacious article and book titles, and statements argue that a description of mind and, with that, self without regard for our bodies and the way they are socially and culturally embedded and shaped by the technological environment beyond our skin and skull is simply a fantasy.[68] We are ourselves because we are "not locked up in a prison of our own ideas and sensations" but rather "environmentally plugged in" to our surroundings.[69]

An even more radical understanding of the self in relationship to sensing machines lies in a philosophical position called the *extended mind thesis*. Rather than focus on a brain cut off from the world, philosophers and scientists with an interest in the extended mind ask, "Where does the mind stop and the rest of the world begin?" The extended mind argues that the environment we find ourselves in is key to driving our cognitive processes: how we remember, how we solve problems, or how we relate to ourselves and others. Our own bodily actions incorporate and depend on tools (like notebooks or computers) that lie outside of us and that help us make decisions or take actions in specific circumstances. These cognitive aids are not only in our heads but also part of the environment outside of us, thus extending our cognition—that is, our minds—into means, techniques, and tools. In this way, "we actively engage a world."[70]

In the early 2000s, philosopher Andy Clark, along with cognitive scientist David Chalmers, initially proposed this concept. They provocatively argued that we are "creatures whose minds are special precisely because they are tailor made for multiple mergers and coalitions."[71] We are what Clark calls *natural-born cyborgs*: mergers of humans and machines, but not because we can implant electronic chips and sensors into our skin or grow new senses through our tightly coupled collaboration with sensors and computers.

As humans, we are also technological because we have always adapted our actions and selves based on the tools and systems that surround us, "forever ready to merge our mental activities with the operations of pen, paper, and electronics" in order to understand and move through the world. Sensing machines, those couplings between sensing and computing technologies, brains, and bodies, help in expanding the ongoing link between us as agents and that "nonbiological matrix of machines, tools, props, codes, and semi-intelligent daily objects" that constitutes the world beyond.[72]

Sensing machines are in fact what researchers call *co-constitutive*. They are mutually interlocked with and codependent on us. Despite the elegant simplicity of "self-knowledge through numbers" posed by quantified self-type thinking or the belief that the statistically realized numbers and graphs being displayed on the smartphone are "you," once our minds, selves, and machines are no longer seen as bounded by the skin but part of the larger plugged-in environment, we might begin to understand the "self" not simply as a static, bounded object standing by and waiting to be sensed: neither a simple target for data gathering nor a mere outcome that sensors capture and mathematical models materialize.

Instead, sensing machines that seek to display to us our selves might suggest the possibility of what Clark calls a *soft self*, a "distributed decentralized coalition" of neural, bodily, and technological processes and actions. Aided by but not taken over by sensing machines, the self is an always evolving entity that can form ever-new circuits with the world.[73] The brain cannot thus be isolated from the ever-changing environmental context we find ourselves, and the self cannot simply be seen as embodied in numbers derived from sensor readouts. Like the fluctuations of brainwaves themselves, the self emerges and dynamically changes in time, subjected to the noise of the environment beyond. The brain and body adapt to changes in the environment and, along with this, this self shifts as well. It is part of that ever-changing set of "looping interactions between material brains, material bodies, and complex cultural and technological environments" that we are only partially responsible for steering and controlling.[74]

Epilogue

The year will be 2022 (or later) when you read this. But writing in late 2020, in the midst of an ever-swelling global pandemic, a locked down Europe, a divided postelection United States, and an ever accelerating environmental crisis, imagining that future is a blur. In a small way, the fallout from the worldwide pandemic may have begun to bring about a change in how we perceive the relationships among the natural, social, technical, and cultural environments we inhabit and perhaps even a new understanding of how we as humans are never separate from those environments but symbiotically linked to them in ways we still cannot fathom. And, in the midst of this chaos, sensing machines have not only continued to operate but serve an active role in making these new links and feedback even more apparent.

But how do sensing machines reveal and shape these new human-environment connections? Pandemic-led environmental shifts became the stuff of dozens of scientific studies during the 2020 COVID-19 lockdowns. Between January and April 2020, for example, oceanographers working for Ocean Networks Canada and monitoring sound transmission in the coastal waters off Vancouver Island using underwater hydrophones, began noticing dramatic changes in the overall level of ambient noise usually produced by global commercial shipping-related traffic. Not surprisingly, due to a significant reduction of commercial shipping because of decreased demand during the pandemic, the Canadian researchers recorded significant drop-offs in noise pollution.[1]

This study illustrates a novel link between two areas of study that seem to have nothing to do with each other: acoustics and economics. But the sensors and statistical procedures measuring quantitative changes in sound pressure under the sea revealed something else besides the measure of acoustic phenomena. They exposed the economic dynamics of the commercial shipping industry under extreme conditions of reduced activity.

Ships were not the only entities affected in the cascading economic shutdowns during the spring of 2020. Another oft-repeated news story was the effect of COVID-driven environmental changes on animals. Amateur videos flooded the internet with images of goats freely meandering through Welsh towns and packs of wild monkeys conquering deserted plazas and streets in Southeast Asian cities in search of their normal tourist-provided repasts of bananas and potato chips.

Animals also played a starring role in the COVID-19 Bio-Logging Initiative, a global sensor research consortium focused on biologging, where sensors attached to a wide range of avian, marine, and earthbound creatures that cross diverse ecosystems are used to monitor movement and migration patterns, behavior, actions, and physiology in an effort to "uncover the hidden life of animals." Although begun before the pandemic, the consortium aimed at comparing pre- and post-COVID "on animal" sensor data—for example, data from accelerometers and gyroscopes mounted on birds to measure their changing energy expenditures.[2]

Aggregating mass amounts of global sensor data that can track the movement or actions of animals is no simple task. The complexity of scale, fusion of heterogenous data sets, gaps in measurement from interruptions in data capture, and lack of international standards all make the required collection and analysis of mass data difficult and costly and demanding of massive multilateral cooperation. What is evident is how the novel coronavirus, a biological entity of approximately 150 microns that isn't even considered alive because it cannot reproduce without a host, managed to alter the level of human activity and production worldwide so completely that birdsong was suddenly unveiled, traffic noise was reduced, and seismographs, unlike before, can now differentiate between human-produced vibrations and the inner rumblings of the earth.

Researchers gave these pandemic transformations a new name: the *anthropause*, "a considerable global slowing of modern human activities, notably travel."[3] An article in the scientific journal *Nature* put it more forcefully: the pandemic "disruption is unprecedented in the modern era of global observing networks, pervasive sensing and large-scale tracking of human mobility and behaviour, creating a unique test bed for understanding the Earth System."[4]

Epilogue

Human researchers were confined to their homes during on-again, off-again pandemic-spawned lockdowns. But sensors, microprocessors, and software mounted on satellites and rooftops and planted in fields and forests worked overtime, monitoring the earth, the skies, and the atmosphere for new anthropause-related changes while these devices remained blissfully unaware of the rapidly unfolding human crisis. Crop changes, forest fire tracking, air quality shifts, biodiversity measurements, excess heat radiated from the ground, less rain, and altered composition of greenhouse gasses were only a few of the indirect environmental consequences of the coronavirus that automated sensing systems seemed to have picked up.

Scientists took particular advantage of these sensing infrastructures during shutdowns to glean information about changing environmental conditions that would have been impossible to assess without these technologies. For example, ECOSTRESS,[5] a NASA-sponsored thermal imaging–based sensing system on the International Space Station orbiting Earth's thermosphere, was harnessed to record thermal-imaging data during the spring of 2020 to understand whether the San Francisco Bay Area was actually heating up more than normal due the lack of vehicles on roads and empty parking lots that usually serve to absorb and reflect back the sun's rays.[6]

Meanwhile, even with weather forecasting disruptions due to the reduction of flights and accompanying loss of atmospheric sensor data, networks of environmental sensors measuring molecules and wavelengths of gasses still recorded significant changes in air quality across diverse regions of the planet: drops in nitrous oxide (N_2O) and carbon dioxide (CO_2) due to reduced burning of fossil fuels, as well as changes to chemicals in the normally polluted air surrounding mass transportation infrastructures like airports, highway intersections, and shipping ports.

COVID-19's manifestation might be said to have almost created an entire industry of new sensor technologies aimed at detecting the virus's presence and propagation, all the while folding humans into a new technology-led quest to discover the movement and consequences of an invisible entity. The main sensor industry–based journal (*Sensors*) devoted an entire issue in 2021 to technologies designed to "detect and diagnose the new coronavirus." Sensing and analysis systems originally designed for security and antiterrorism applications metamorphosed into virus fighters: sensors originally created for sniffing out illegal drugs and explosives as coronavirus breathalyzers; detectors made out of electronic filaments one hundred times thinner than human hair to test the amount of breath moisture leaked through respiratory masks during normal or rapid breathing or coughing; the application of AI-based speech recognition to monitor coughing and speech to detect viral presences; new optical and thermal biosensors with artificial DNA sequences embedded into gold-based nanostructures that can identify the RNA strands present in SARS-CoV-2 viral strains; or chemical sensors measuring molecular changes in a device's immediate environment to model the aerosol-like spread of the virus in the air or on surfaces on which it might land.[7]

It is perhaps ironic that a group of chemical sensors not electronic in origin were also dramatically affected by the virus. The human chemical senses of smell and taste, the so-called lower senses long ignored by researchers, appeared to be the first human senses radically affected by SARS-CoV-2. Those infected claimed a sudden loss of their ability to smell (*anosmia*) from one moment to the next, only for these senses to return several weeks later, at least for 90 percent of the infected.

The victims of COVID-induced smell loss claimed strange sensations: drinking coffee and no longer being able to taste anything, or suddenly not being able to smell their partner or children. Others affected underwent "physical therapy for the nose," with charities offering "smell therapy" through daily regimens of inhaling essential oils in order to rebuild olfactory cells and their associated neural networks damaged by the virus.[8] The circumstances behind these transformations of smell and taste were equally baffling. Researchers found that the virus altered the smell sense in patients not directly (by infecting the olfactory sensory receptor neurons) but indirectly, by damaging supporting cells that assist the main sensory receptors.[9]

With such startling reports, we might venture that the coronavirus resulted in a new kind of *unmaking of sense* by robbing the human senses of their ability to smell and taste the world. If our human senses were confused, then substitutes—human-made electronic sensors deployed in our surrounding environments that continually track us as part of a new theater of control and containment—stood by. From contact-tracing wristbands to sensors installed in the ceilings of offices to enforce social distancing to a Japanese-developed robot using IR lasers and computer vision to remind customers to "mask up" or keep a safe distance from others, the pandemic condition greatly enlarged machine sensing systems' remit to track, trace,

Epilogue

and constrain human mobility.[10] That these distributed, wearable, ubiquitous, and pervasive monitoring systems all rely on machine intelligence to make sense of human-generated data is hardly a surprise. Unlike human neural networks, which fell prey to the coronavirus, the artificial neurons running machine learning prediction models in the various COVID contact-tracing apps remained resilient against the threat of mere biological contamination.[11]

What is going on here? The long-term effects of what biodiversity researchers have called the "global human confinement experiment" are not yet known.[12] But endlessly proliferating sensors designed to control populations and detect viral particles, ubiquitous in 2020 newsfeeds, suggest that a new kind of sensing imaginary arrived—one partially built on engineering opportunism coupled with the never-wavering belief that technology can be our savior in the face of biomedical and environmental crisis. Machines seem to take over from our human inability to not get too close to each other or to detect lethal molecules propagating through the air.

But there is another side to this admittedly bleak and surreal narrative. This book has demonstrated how sensing machines have become ubiquitous in our daily lives, almost to the point where we don't even notice them anymore. At the same time, I hope to have convinced you that they have not appeared out of the blue. Their ability to sense, monitor, and change us and the environments we inhabit has deep historical roots and multiple contexts and motivations that are not only the by-product of surveillance capitalism, propagated by a few powerful, late twentieth-century information age corporations.[13]

Indeed, many (but not all) of the sensing machines that we have seen in these pages are grounded in a technological worldview that although never neutral is also not just predatory. In other words, the imaginaries harbored for sensors and machine intelligence that thoroughly reconfigure our lived experience of play, transport, art, food, health, dreams, and selfhood suggest multiple futures—fantastic, dystopian, speculative, ludicrous, and pragmatic.

The pandemic that raged across the world in 2020–2021 and perhaps even still reveals something else: a heightened awareness of the roles sensing machines play in creating new encounters among us and our technological and natural environment. Why is this?

First, these encounters suggest the further expansion of an ecological approach to understanding how our sensor technology, our perception, and our environment form an inseparable, holistic, interdependent whole. The term *ecological* here has perhaps a slightly different meaning than as we normally understand it. According to American experimental psychologist James J. Gibson in the 1960s, who developed an "ecological theory of perception," ecological describes a *reciprocity* between perceiver and environment: "The ecological approach takes as its unit of study the animal in its environment, considered as an interactive system. The relations within this system are reciprocal, with the reciprocity including a species evolving in an environment to which it becomes adapted, and an individual acting in its own niche, developing and learning."[14]

Gibson's ecological theory suggests a symbiosis, a living together of different organisms, between the perceptual apparatus and the environment that it evolved to perceive—a worldview in which organism and environment are inseparable from each other. Sensing and acting are one. Humans and their environments, natural, social, and technological, are mutually dependent on each other. Our actions affect not only our own perception but the perception inherent in the worlds surrounding us. If you doubt this, just look at how the different transmission styles of the coronavirus emphasize such reciprocality: environment to human, human to human, animal to human, human to other.[15]

At the same time, as we scrambled in 2020–2021 to use sensors and machine intelligence to track and trace something wholly invisible to our naked eye in order to prevent its spread across humans, animals, objects, and the very air we breathe, sensing machines' capacity to entangle different natural and technological entities moved front and center. Sensors made of electronics, cells, and chemicals attempting to sense traces of a virus inside laboratories. Wearable sensors on human arms and embedded software in phones aiming to trace social connections. Sensors under the sea and in the ground measuring unheard vibrations from the earth for the first time due to the lack of overwhelming human-generated vibrations that usually mask these sounds. Sensors mounted on flights no longer producing accurate renderings of temperature, wind speed, or barometric pressure as there were simply not enough of these machines in the air at the same time.

All of these entities—half biological, half technical, social, and atmospheric—embody phenomena that scholars call *hybrids*—new combinations of social, natural, and technological things, objects, and species that defy our tried and true (but very worn out) ways of understanding the complex relationship between us and the social-natural-technical world we inhabit.[16] Hybrids are definitely worth paying attention to. They fundamentally challenge us to think about how sensing and sense-making capacities of different entities—human, machine, animal, plant, cell—inseparably shape, affect, and change each other.

In chapter 5, we met the Austrian economist and political philosopher Friedrich A. Hayek, whose 1952 book *The Sensory Order* not only set out early ideas of the concept of neural networks but also inspired a contemporary scientist (Takashi Ikegami) to create a sensing machine in the form of android that through its sensors aimed to develop an artificial mind.

Hayek's concept of the sensory order is appropriate to our (post-)pandemic world, especially in light of an increasing awareness of the deep entanglement between humans, machines, and their environments. As Hayek argued, the *sensory order* is a *product of the nervous system*. The smell of freshly baked bread, the sound of the pounding ocean, or the blinding dazzle of sunlight against a snow-dusted landscape is the result of a complex symphony of billions of interconnected neurons continually organizing and reorganizing and firing electrical and chemical signals between each other.

To uncover the relationship between stimuli, sensation, and perception, Hayek needed to show "why and how the senses classify similar physical stimuli sometimes as alike and sometimes as different."[17] These structures of difference are what Hayek called *orders*: an arrangement of parts of a greater whole put in relationship to one another according to a preconceived plan. But Hayek's *sensory order* is different. It is not planned or humanly designed but emerges in a spontaneous and self-organizing way because it is an "order which is not made by anybody but which forms itself."[18] Perception is an ordering system.

Hayek's reimagination of sensation and perception in a vastly similar way to the widespread explosion of machine-based deep learning at the end of the 2000s has an air of the uncanny to it. Sensation and perception are recast as emergent processes, in both humans and machines. The brain itself does not sense and make meaning from top-down rules and symbolic structures but rather from spontaneous, bottom-up connections, reconnections, and continual configuring of neurons, based on a change of connections or weights between neurons in order to learn and "experience" the world.

Quite simply, in our computational machine age, sensing is redefined. It is not simply the mind looking for physical correspondences "out there" in the world. Rather, the sensory order in machine-based sensing and learning depends on former experiences; its operations are spontaneous, emergent, complex, and beyond our human knowing. Even the scientists building these new human-machine configurations don't fully know what is going on, fulfilling Hayek's belief about the limits of knowledge that we can have concerning any complex, spontaneously arising order.

Less than a century before Friedrich Hayek, Gustav Fechner, whom we began this book with, also sought to understand the relationship between the worlds of physical phenomena and sensory experience. Seeking to heal the split between mind and body, the physical and psychic universes, Fechner reached for the unbridled power of mathematics to unlock the mysteries of stimuli, sensation, perception, and experience.

For Fechner, sensing machines were still human beings. Hayek takes the next step. Even though *The Sensory Order* is ostensibly about the human mind, more than human minds haunt its pages. In fact, in more than a few examples, Hayek describes a machine with the task to classify and order balls of various sizes and distribute them. This thought machine sorts out objects. It is, for all intents and purposes, a toy designed to communicate how the mind classifies an undifferentiated mass of stimuli coming at it.

Hayek also describes another machine. This one sorts out "individual signals arriving through any one of a large number of wires or tubes." This sorting and classifying machine, has a striking similarity to another kind of "mind" that might not qualify as human. In fact, it is more like a computer, in which "certain statistical machines for sorting cards on which punched holes represent statistical data," Hayek wrote, and "if we regard the appearance of any card with the same data punched on it as the recurrence of the same event, and assume that the machine is so arranged that various groups of different data are placed into the same receptacle, we should then have a machine which performs a classification in the sense in which we use this term."[19]

By specifying a machine that classifies the noisy stimuli of the world into new sensory orders, Hayek has the upper hand as he anticipates what is to come. It is not by chance that a mere six years after *The Sensory Order*'s publication, a Cornell University psychologist named Frank Rosenblatt implemented the first classifying neural network, called the Perceptron, directly in the physical guts of computer hardware.[20] Channeling Hayek, whom Rosenblatt cites in his groundbreaking article for *Psychology Review*, "The Perceptron: A Probabilistic Model for Information Storage and Organization in the Brain," Rosenblatt's thoughts made materialized in a computational machine confronted three riddle of the ages: "(1) How is information about the physical world sensed or detected by the biological system; (2) In what form is information stored or remembered, and (3) How does information contained in storage, or in memory, influence recognition and behavior?"[21]

And now, in the twenty-first century, sensing machines have come full throttle out of the shadows. Configured like Fechner's quantified bodies and Hayek's neurally wired, reconfiguring minds, our sensing, classifying, and learning machines are so integrated in how we now live and breathe that there is less and less separation between us and them anymore. But as this book has tried to show, seen from the imaginaries of scientists, engineers, artists, designers, architects, and technologists, this separation was fragile in the first place. Our senses and selves have long ceased to be "other" from the technical world (if they ever were). Sensing machines are not simply "out there." They are us.

Acknowledgments

For an author, acknowledgements are the most difficult part of the book to write because there are always so many people who have helped bring a new work into existence. And it's inevitable that someone always gets left out. So, the easiest option is just to include an alphabetical list with a few special thanks at the end: Erik Adigard, Marie-Luise Angerer, Sofian Audry, Ars Electronica, Baltan Labs, Barbican Centre, Josh Berson, Jennifer Biddle, Peter Cariani, Jadwiga Charzyńska, Jean Dubois, Karmen Franinović, Orit Halpern, Jens Hauser, David Howes, Takashi Ikegami, Sidd Khajuria, Laznia Center for Contemporary Art, Garrett Lockhart, Claudia Mareis, Philip Mirowski, Marie Morin, David Parisi, Simon Penny, Josep Perelló, RIXC, Joel Ryan, Alex Saunier, Henning Schmidgen, Silke Schmidt, Bart Simon, TeZ, Joseph Thibodeau, Jose Luis de Vincente, Marcelo Wanderley, and Arnd Wesemann. Special thanks to my anonymous reviewers at the MIT Press; my multitude of students and colleagues; the Office of the Vice President of Research at Concordia University for five years of support from the Concordia University Research Chair program; Viktoria Tkaczyk and the Max Planck Institute for the History of Science in Berlin; my former long-time editor at MIT Press, Doug Sery, for believing in all my books; Noah Springer, Kathleen Caruso, Melinda Rankin, and the rest of the (ever) excellent MIT Press editorial and design team and, as always, Anke. Some of the artistic projects described in these pages were supported by grants from the Fonds de Recherche du Québec, Société et Culture, and the Social Science and Humanities Research Council of Canada. This book is dedicated to the memory of David Patch, Geoffrey Reeves, and Carl Weber, three of my artistic-scholarly mentors who passed away during the six years it took me to write it and who taught me that critical reflection and imaginative creation go hand in hand.

Notes

Prologue

1. This introduction is loosely based on the following already existing or currently research-based technologies, scenarios, products, or applications: Jessica Zimmer, "Fighting COVID-19 with Disinfecting Drones and Thermal Sensors," Engineering.com, June 12, 2020, https://new.engineering.com/story/fighting-covid-19-with-disinfecting-drones-and-thermal-sensors; "Beware the IoT Spy in Your Office or Home via Smart Furniture, Warns the NSA," CSO, October 31, 2018, https://www.csoonline.com/article/3317938/beware-the-iot-spy-in-your-office-or-home-via-smart-furniture-warns-nsa.html; Sidney Fussell, "The City of the Future Is a Data-Collection Machine," *Atlantic*, November 21, 2018, https://www.theatlantic.com/technology/archive/2018/11/google-sidewalk-labs/575551; Michael W. Sjoding, Robert P. Dickson, Theodore J. Iwashyna, Steven E. Gay, and Thomas S. Valley, "Racial Bias in Pulse Oximetry Measurement," *New England Journal of Medicine* 383:2477–2478; Adam Carter and John Rieti, "Sidewalk Labs Cancels Plan to Build High-Tech Neighborhood in Toronto amid COVID-19," CBC, May 7, 2020, https://www.cbc.ca/news/canada/toronto/sidewalk-labs-cancels-project-1.5559370; Horatiu Boeriu, "BMW Natural Interaction Introduced at the Mobile World Congress 2019," BMWBLOG, February 25, 2019, https://www.bmwblog.com/2019/02/25/bmw-natural-interaction-introduced-at-the-mobile-world-congress-2019; "This Is CogniPoint," PointGrab, accessed September 5, 2020, https://www.pointgrab.com/our-product; Bin Yu, Mathias Funk, and Loe Feijs, "DeLight: Biofeedback through Ambient Light for Stress Intervention and Relaxation Assistance," *Personal and Ubiquitous Computing* 22, no. 4 (2018): 787–805, https://link.springer.com/article/10.1007/s00779-018-1141-6; Stacey Cowley, "Banks and Retailers Are Tracking How You Type, Swipe and Tap," *New York Times*, August 13, 2018, https://www.nytimes.com/2018/08/13/business/behavioral-biometrics-banks-security.html; Sarah Mitroff, "Hitting the Pavement with Spotify Running," CNET, June 13, 2015, https://www.cnet.com/news/hitting-the-pavement-with-spotify-running-hands-on; Philip Qian and Esge B. Andersen, Earbuds with Biometric Sensing, US Patent 9716937B2, filed September 16, 2015,

and issued July 25, 2017; https://www.teamlab.art, teamLab, accessed November 20, 2020; and https://elbarri.com/en, elBarri, accessed November 15, 2020.

2. "IoT Sensors and Actuators," SBIR/STTR, accessed November 20, 2020, https://www.sbir.gov/node/1319475. It's not just the number of sensors that is skyrocketing. It's also how much data these sensors are generating, with estimates somewhere in the forty zettabyte range, or forty trillion gigabytes.

3. Lucia Maffei, "Boston-Made Fitness Tracker Is Being Used to Track COVID-19 Symptoms," *Boston Business Journal*, March 23, 2020, https://www.bizjournals.com/boston/news/2020/03/23/boston-made-fitness-tracker-is-being-used-to-track.html; https://www.tracesafe.io.

4. Joellen Russell, "Ocean Sensors Can Track Progress on Climate Goals," *Nature* 555, no. 7696 (2018): 287, https://www.nature.com/articles/d41586-018-03068-w.

5. Geri Piazza, "Spongy Stomach Sensor that Could Be Swallowed," NIH Research Matters, February 26, 2019, https://www.nih.gov/news-events/nih-research-matters/spongy-stomach-sensor-could-be-swallowed.

6. While the notion of cyborgs evokes science fiction stories, the original concept of a *cybernetic organism* derives from the work of Austrian American mathematician, musician, and inventor Manfred E. Clynes and clinical psychiatrist Nathan S. Cline. See Manfred E. Clynes and Nathan S. Kline, "Cyborgs and Space," *Astronautics* 5, no. 9 (1960): 26–27, 74–76.

7. It was also widely reported that drivers are required to wear heartrate monitors, and the bridge employs "yawn cameras" to detect if a driver is on the threshold of falling asleep. See Kate Lyons, "'Yawn Cams' and Heart Monitors: Five Key Facts about the World's Longest Sea Bridge," *Guardian*, October 23, 2018, https://www.theguardian.com/world/2018/oct/23/five-things-you-need-to-know-about-the-worlds-longest-sea-bridge. See also "Want to See Magical Technology behind the Hong Kong-Zhuhai-Macao Bridge? Follow Soway," Soway Tech Limited, May 29, 2019, http://www.sowaytech.com/sdp/302911/4/nd-5117145/184265/News.html.

8. "Inside the Equinix NY4 Financial Trading Hub," Data Center Knowledge, October 14, 2013, https://www.datacenterknowledge.com/inside-the-equinix-ny4-financial-trading-hub.

9. "Empowering Smart Buildings with a 'Digital Brain,'" Arup, accessed April 15, 2020, https://www.arup.com/projects/neuron.

10. Stephen Chen, "'Forget the Facebook Leak': China Is Mining Data Directly from Workers' Brains on an Industrial Scale," *South China Morning Post*, April 29, 2018, https://www.scmp.com/news/china/society/article/2143899/forget-facebook-leak-china-mining-data-directly-workers-brains.

11. See http://biocatch.com.

12. See David Dennis Jr., "AI Lacks Intelligence without Different Voices," x.ai.com, May 9, 2018, https://x.ai/ai-lacks-intelligence-without-different-voices/; Joy Buolamwini and Timnit Gebru, "Gender Shades: Intersectional Accuracy Disparities in Commercial Gender Classification," Proceedings of the 1st Conference on Fairness, Accountability and Transparency, *PMLR* 81 (2018):77–91; and Halcyon Lawrence, "Siri Disciplines," in *Your Computer Is on Fire!*, ed. T. S. Mullaney, B. Peters, M. Hicks, and K. Philip (Cambridge, MA: MIT Press, 2021), 179–198.

13. Caroline Ku, "Airplane Seat that Monitors Heart Rate Could Also Save Airlines Money," APEX, May 14, 2015, https://apex.aero/articles/airplane-seat-that-monitors-heart-rate-could-also-save-airlines-money/.

14. See David Howes, "Hyperesthesia, or, the Sensual Logic of Late Capitalism," in *Empire of the Senses: The Sensual Culture Reader* (Oxford: Berg, 2005), 281–303.

15. See Theodore M. Porter, *The Rise of Statistical Thinking, 1820–1900* (Princeton, NJ: Princeton University Press, 2020); and Stephen M. Stigler, *The History of Statistics: The Measurement of Uncertainty before 1900* (Cambridge, MA: Harvard University Press, 1986).

16. Ian Hacking, "Making Up People," in *Reconstructing Individualism*, ed. T. L. Heller, M. Sosna, and D. E. Wellbery (Stanford, CA: Stanford University Press, 1986), 222–236. See also Michel Foucault, *Security, Territory, Population: Lectures at the Collège de France, 1977–78*, ed. Michael Senellart, trans. Graham Burchell (New York: Picador, 2009); Michel Foucault, *Discipline and Punish: The Birth of the Prison*, trans. Alan Sheridan (New York: Vintage, 1979).

17. Shoshana Zuboff, *The Age of Surveillance Capitalism* (New York: Public Affairs, 2019); Deborah Lupton, *The Quantified Self* (Cambridge: Polity, 2016); S. D. Esposti, "When Big Data Meets Dataveillance: The Hidden Side of Analytics," *Surveillance & Society* 12, no. 2 (2014): 209–225; Nigel Thrift, *Knowing Capitalism* (London: Sage, 2005); Deborah Lupton, *Data Selves: More-than-Human Perspectives* (Cambridge: Polity, 2020); Rob Kitchin, "Big Data, New Epistemologies and Paradigm Shifts," *Big Data & Society* 1, no. 1 (2014): 1–12; and Mark Andrejevic, *iSpy: Surveillance and Power in the Interactive Era* (Lawrence: University of Kansas Press, 2007).

18. See Zuboff, *The Age of Surveillance Capitalism*.

19. See Lee Rainie and Janna Anderson, "The Future of Privacy: Above-and-Beyond Responses: Part 1," Pew Research Center, December 18, 2014, https://www.pewresearch.org/internet/2014/12/18/above-and-beyond-responses-part-1-2.

20. See Sundar Sarukkai, "Praying to Machines," *Leonardo Electronic Almanac* 11, no. 8 (August 2003), https://www.leoalmanac.org/leonardo-electronic-almanac-volume-11-no-8-august-2003/; and P. Hill, *The Book of Knowledge of Ingenious Mechanical Devices (Kitāb fī ma'rifat al-ḥiyal al-handasiyya)* (Berlin: Springer Science & Business Media, 2012).

21. See Chris Salter, "Just Noticeable Difference: Ontogenesis, Performativity, and the Perceptual Gap," in *Perception and Agency in Spaces of Contemporary Art*, ed. Christina Albu and Dawna Schuld (London: Routledge, 2018). See also Chris Salter, *Just Noticeable Difference (JND)*, art installation, http://www.chrissalter.com/just-noticeable-difference-jnd.

22. For some examples of scientists, cultural and art historians, and anthropologists who have covered these histories of sensing and quantification, see, among others, Anson Rabinbach, *The Human Motor: Energy, Fatigue, and the Origins of Modernity* (Berkeley: University of California Press, 1992); Robert Brain, *The Pulse of Modernism: Physiological Aesthetics in Fin-de-Siècle Europe* (Seattle: University of Washington Press, 2015); Henning Schmidgen, *The Helmholtz Curves: Tracing Lost Time*, trans. Nils Schott (New York: Fordham University Press, 2014); Jimena Canales, *A Tenth of a Second: A History* (Chicago: University of Chicago Press, 2011); Kurt Danziger, *Constructing the Subject: Historical Origins of Psychological Research* (Cambridge: Cambridge University Press, 1990); and Jonathan Crary, *Techniques of the Observer: On Vision and Modernity in the Nineteenth Century* (Cambridge, MA: MIT Press, 1990). For a more recent anthropological take, see Josh Berson, *Computable Bodies: Instrumented Life and the Human Somatic Niche* (London: Bloomsbury Publishing, 2015); and Joseph Dumit and Marianne de Laet, "Curves to Bodies," in *Routledge Handbook of Science, Technology and Society*, ed. Daniel Lee Kleinman and Kelly Moore (London: Routledge, 2014), 71–90.

Chapter 1

1. The account of Gustav Fechner's malady comes from his biography. See Johannes Emilie Kuntze, *Gustav Theodor Fechner. Ein deutsches Gelehrtenleben* (Leipzig: Breitkopft und Härtel, 1892). This has been excerpted in an English-language edition: *Religion of a Scientist: Selections from Gustav Th. Fechner*, ed. and trans. Water Lowrie (New York: Pantheon, 1946), 36–42.

2. Fechner, *Religion*, 36–37.

3. Kuntze, *Ein deutsches Gelehrtenleben*, 108. My translation.

4. Fechner, *Religion*, 41.

5. "Research Scientist—Neural Interfaces," Facebook, October 9, 2020, job posting, https://www.mendeley.com/careers/job/research-scientist-neural-interfaces-700907.

6. "Research Scientist, Applied Perception Science: AR/VR," Oculus, accessed April 10, 2020, https://www.oculus.com/careers/a1K2K000007stPKUAY.

7. Gustav Fechner, *Elements of Psychophysics: Volume 1*, ed. Edwin G. Boring and Davis H. Howes, trans. Helmut E. Adler (New York: Holt, Rinehart and Winston, 1966), xxvii.

8. For a critique of psychophysics, see Friedrich Kittler, "Thinking Colours and/or Machines," *Theory, Culture & Society* 23, no. 7–8 (2006): 39–50.

9. François Dagognet, *Étienne-Jules Marey: A Passion for the Trace*, trans. Robert Galeta and Jeanine Herman (New York: Zone Books, 1992), 15

10. See David Skrbina, *Panpsychism in the West* (Cambridge, MA: MIT Press, 2005).

11. Fechner, *Elements*, xxiv.

12. For more detail, see David Parisi, *Archaeologies of Touch: Interfacing with Haptics from Electricity to Computing* (Minneapolis: University of Minnesota Press, 2018).

13. See Edwin G. Boring, *Sensation and Perception in the History of Experimental Psychology* (New York: Appleton-Century-Crofts, 1942), 37–40.

14. The equation was as follows: S (sensation) = ΔI (the minimum change of the stimulus's intensity necessary for the experience of the jnd) / I (the intensity of the stimulus) * k (a constant).

15. For those readers with a mathematical interest, Fechner ends up with a logarithm (logR) by expressing the equation S = ΔI / I * k in differential equation terms. A *differential equation* is one that involves variables that represent the rates of change of continuing varying quantities (called *derivatives*). Fechner's equation thus rewrote Weber's law in the following way: dS (the change of sensation) = k * dI / I. To solve the equation, you have to integrate and end up with the aforementioned solution, in which sensation is a logarithmic function of the intensity of the stimulus.

16. In the case of both Ohm's law and Fechner's law, logarithms are used to calculate large numbers. As something increases by one unit, logarithms indicate that the effect is increased by ten times, and then the result increases by a factor of ten.

17. See Crary, *Techniques*, 148, and Rabinbach, *Human Motor*, for the link between psychophysics and the conservation of energy.

18. Edwin G. Boring, *A History of Experimental Psychology* (Englewood Cliffs, NJ: Prentice Hall, 1957), 280.

19. Fechner, *Elements*, 60–62.

20. The *method of limits* involves presenting a stimulus to the subject in ascending or descending levels of intensity to determine the threshold where the smallest amount becomes detectable. The *method of adjustment* involves the continuous adjustment of the intensity of a stimulus until the subject can or cannot perceive it. The *method of constant stimuli* or *right and wrong cases* presents a stimulus in a random order, thus preventing the subject from predicting the next intensity level of the stimulus. Fechner, *Elements*, 61–111.

21. Chiao Liu, Michael Hall, Renzo De Nardi, Nicholas Trail, and Richard Newcombe, "Sensors for Future VR Applications" (paper presented at the International Image Sensor Workshop, Hiroshima, Japan, May 30–June 2, 2017).

22. Hugh Langley, "Inside-out v. Outside-in: How VR Tracking Works, and How It's Going to Change," Wareable, May 3, 2017, https://www.wareable.com/vr/inside-out-vs-outside-in-vr-tracking-343.

23. Beatrice de Gelder, Jari Kätsyri, and Aline W. de Borst, "Virtual Reality and the New Psychophysics," *British Journal of Psychology* 109, no. 3 (2018): 421–426.

24. See C. Tilikete and A. Vighetto, "Oscillopsia: Causes and Management," *Current Opinion in Neurology* 24, no. 1 (2011): 38–43.

25. Bernard D. Adelstein, Thomas G. Lee, and Stephen R. Ellis, "Head Tracking Latency in Virtual Environments: Psychophysics and a Model," *Proceedings of the Human Factors and Ergonomics Society Annual Meeting* 47, no. 20 (2003): 2083–2087.

26. Qi Sun, Anjul Patney, Li-Yi Wei, Omer Shapira, Jingwan Lu, Paul Asente, Suwen Zhu, Morgan Mcguire, David Luebke, and Arie Kaufman, "Towards Virtual Reality Infinite Walking: Dynamic Saccadic Redirection," *ACM Transactions on Graphics* 37, no. 4 (2018): 1–13.

27. Sun et al., "Towards Virtual Reality," 2.

28. Alex Wawro, "Inside Magic Leap: How It Works and What It Means for Game Devs," Gamasutra, August 8. 2018, https://www.gamasutra.com/view/news/323455/Inside_Magic_Leap_How_it_works_and_what_it_means_for_game_devs.php.

29. "Magic Leap One Teardown," iFixit, August 23, 2018, https://www.ifixit.com/Teardown/Magic+Leap+One+Teardown/112245.

30. See, for example, Sylvain Chagué and Caecilia Charbonnier, "Real Virtuality: A Multi-user Immersive Platform Connecting Real and Virtual Worlds," *Proceedings of the 2016 Virtual Reality International Conference* (Laval, France: ACM Press, 2016), 1–3.

31. This list includes some of the many instruments developed at the time for psychology and perception research. See Brain, *Pulse of Modernism* and Dagognet, *Étienne-Jules Marey* for more details.

32. See Brain, *Pulse of Modernism*, 48–49.

33. Dagognet, *Étienne-Jules Marey*, 61.

34. See, for example, Schmidgen, *Helmholtz Curves*, and Canales, *Tenth of a Second*, for more on Helmholtz's time experiments.

35. See Brain, *Pulse of Modernism*, 72–136, for an exhaustive study.

36. See Étienne-Jules Marey, *La méthode graphique dans les sciences expérimentales et principalement en physiologie et en médecine* (Paris: G. Masson, 1885).

37. See Edward R. Tufte and Peter R. Graves-Morris, *The Visual Display of Quantitative Information* (Cheshire, CT: Graphics Press, 1983).

38. See Nolwenn Maudet, "Muriel Cooper-Information Landscapes," accessed June 4, 2021, http://www.revue-backoffice.com/en/issues/01-making-do-making-with/nolwenn-maudet-muriel-cooper-information-landscapes.

39. Experimenting with sensory perception was not the only aim of Wundt's research. In fact, it has been argued that this work was minor compared to Wundt's larger interest in what is called in German *Vorstellung*—literally, perceptions and ideas. See Kurt Danziger, "Wilhelm Wundt and the Emergence of Experimental Psychology," in *Companion to the History of Modern Science*, ed. R. C. Olby, C. N. Cantor, J. R. R. Christie, and M. J. S. Hodge (London: Routledge, 1990), 396–409.

40. See Henning Schmidgen, "Camera Silenta: Time Experiments, Media Networks, and the Experience of Organlessness," *Osiris* 28, no. 1 (2013): 162–188.

41. See Wilhelm Wundt, "Das Institut für experimentelle Psychologie zu Leipzig," in *Psychologische Studien*, vol. 5, no. 6 (Leipzig: Wilhelm Engelmann, 1907), 279–293.

42. These researchers from the United States (some 33 in total), France, England, Belgium and other countries sought not only to study and learn from Wundt's extensive battery of experimental procedures and methods but also to return to their own countries to found or strengthen their own laboratories at home. Tellingly, Wundt graduated only one female student—the psychologist Anna Berliner, who later went on to do pioneering work in the United States on the relationship between psychology and culture. For more on Berliner, see Rachel Uffelman, "Anna Berliner (1888–1977)," *The Feminist Psychologist* (Newsletter of the Society for the Psychology of Women, Division 35 of the American Psychological Association) 29, no. 2 (Spring 2002), https://www.apadivisions.org/division-35/about/heritage/anna-berliner-biography.

43. See Danziger, *Constructing the Subject*, 31.

44. See Edward B. Titchener, *A Text-Book of Psychology* (New York: Macmillan, 1928), 246.

45. See Simon Schaffer, *From Physics to Anthropology and Back Again* (Cambridge: Prickly Pear Press, 1994), 22.

46. Stefano Sandrone, Marco Bacigaluppi, Marco R. Galloni, Stefano F. Cappa, Andrea Moro, Marco Catani, Massimo Fillippi, Martin M. Monti, Daniela Perani, and Gianvito Martino, "Weighing Brain Activity with the Balance: Angelo Mosso's Original Manuscripts Come to Light," *Brain* 137, no. 2 (2014): 621–633.

47. Daniel J. Cuthbert, User Identification System Based on Plethysmography, US Patent 20160296142, filed December 30, 2013, and issued October 13, 2016.

48. C. Régnier, "Étienne-Jules Marey, the 'Engineer of Life,'" *Medicographia* 25 (2003): 268–274.

49. Dagognet, *Étienne-Jules Marey*, 30.

50. See Canales, *Tenth of a Second*, 71.

51. The graphic method was indeed gradually abandoned by Marey with his development of chronophotographic methods—methods that provided a much more accurate trace of movement.

52. See Canales, *Tenth of a Second*. See also Andreas Mayer, "The Physiological Circus: Knowing, Representing, and Training Horses in Motion in Nineteenth-Century France," *Representations* 111, no. 1 (2010): 88–120.

53. This covers both hardware sensing issues as well as software. For a more detailed and clear explanation of bias across all of the stages of the machine learning cycle, see Ayesha Bajwa, "What We Talk About When We Talk About Bias (A guide for everyone)," Medium.com, August 18, 2018, https://medium.com/@ayesharbajwa/what-we-talk-about-when-we-talk-about-bias-a-guide-for-everyone-3af55b85dcdc.

54. Ben Court, "Inside Apple's Secret Performance Lab," *Men's Health*, February 2, 2017, https://www.menshealth.com/technology-gear/a18923364/inside-apples-secret-performance-lab.

55. A student of the famous French biologist Claude Bernard would later develop a portable chronoscope, a central instrument used to conduct reaction time experiments. This would actually enable that expensive and fragile device to leave the safety of the laboratory and enter into the field.

56. *Window* is a technical term that comes from digital signal processing. It describes a smaller subset, a window, of a larger continuous signal.

57. Two such techniques frequently used in smartphones and fitness trackers are autocorrelation and autoregression, which measure and predict future actions based on past actions. *Autocorrelation* shows the degree of similarity or correlation between the values of the same variables over successive time intervals, while *autoregression* uses past values of a variable in a time series to make predictions about future values of those variables.

Chapter 2

1. Joel Ryan, "Effort and Expression," in *Proceedings of the 1992 International Computer Music Conference* (San Jose, CA: Computer Music Association, 1992), 414–416.

2. See Steven Spier, *William Forsythe and the Practice of Choreography: It Starts from Any Point* (London: Routledge, 2011).

3. Roberto Calasso, *The Marriage of Cadmus and Harmony*, trans. Tim Parks (Toronto: Vintage, 1994).

4. STEIM officially closed in 2020 after forty-nine years in existence. For an early history, see Joel Ryan, "Some Remarks on Musical Instrument Design at STEIM," *Contemporary Music Review* 6, no 1 (1991): 3–17; and Chris Salter, *Entangled: Technology and the Transformation of Performance* (Cambridge, MA: MIT Press, 2010), 204–205.

5. Curtis Roads, *The Computer Music Tutorial* (Cambridge, MA: MIT Press, 1996).

6. Joel Ryan in discussion with the author, May 2020.

7. For a discussion of the differences between the large computer music research centers like Stanford in contrast to the musician-driven context of Mills College, see Douglas Kahn, "Between a Bach and a Bard Place," in *Media Art Histories*, ed. Oliver Grau (Cambridge, MA: MIT Press, 2007), 423–451.

8. The earliest known version of the apparatus seems to derive from the eighteenth century, developed by the English physicist Reverend George Atwood. Atwood aimed to demonstrate Sir Isaac Newton's breakthroughs in translating the actions of the physical universe—and in particular, the relationship between gravity and acceleration—into mathematical formulae by designing a contraption consisting of two weights connected by a massless string running over a pulley. This device exposed the direct relationship of mass, acceleration, and gravity. At around the same time, pendulum-based clocks were also used to measure acceleration by way of gravity by calculating the relationship between a pendulum's length and the time it took to swing through one cycle or period. Bulky and heavy, these pendulums were not easily attached to moving objects (e.g., they were essentially useless on ships due to rocking caused by waves).

9. McCollum and Peters were interested in calculating the short but disastrous vibrations called *transients*—high-frequency waves that move through material like steel and can produce volatile stresses and strains in *structural members*, the supports that keep buildings, bridges, machines, and other large-scale mechanical systems from collapsing. See Burton McCollum and Orville Sherwin Peters, "A New Electric Telemeter," paper no. 247 (Washington, DC: National Bureau of Standards, 1924); and Patrick Walter, "Review: Fifty Years plus of Accelerometer History for Shock and Vibration," *Shock and Vibration* 6, no. 4 (1999): 197–207.

10. USGS, "Strong Motion," Earthquake Glossary, accessed October 20, 2019, https://earthquake.usgs.gov/learn/glossary/?term=strong%20motion.

11. David Bressan, "Nikolai Tesla's Earthquake Machine," *Forbes*, January 7, 2020, https://www.forbes.com/sites/davidbressan/2020/01/07/nikola-teslas-earthquake-machine.

12. Deyan Sudjic, "At Last—a Bridge You Can Cross," *Guardian*, March 10, 2001, https://www.theguardian.com/theobserver/2001/mar/11/2.

13. Andy Beckett, "Shaken Not Sturdy," *Guardian*, July 18, 2000, https://www.theguardian.com/artanddesign/2000/jul/18/architecture.artsfeatures.

14. To give a sense of the range of g-forces, normally standing on the earth is based on a g-force of 1 g. As the g-force builds up, the force pushes blood away from the heart and toward the legs, making it increasingly difficult for the blood to recirculate back to the heart and to the brain. A g-force of 3 g can deprive the brain of oxygen.

15. Richard P. Feynman, "There's Plenty of Room at the Bottom," *Resonance* 16, no. 9 (2011): 890–905.

16. Feynman, "Plenty of Room," 898.

17. In the mid-1990s, Ryan collaborated with the pioneering video artist Steina Vasulka, who used her amplified violin to control a Pioneer video laser disc player equipped with a serial interface to allow serial-based electronics (like the STEIM SensorLab) to be connected to it and to control the speed, forward and backward play, and pause on the device. Vasulka, who had bought a cache of these laser disc players from Hollywood film studios who had used them for editing before computer editing became popular, had brought these laser disc players to STEIM. Ryan replaced the violin with an accelerometer. See "Steina Vasulka: Violin Power," Digital Canon, accessed November 29, 2020, https://www.digitalcanon.nl/?artworks=steina-vasulka-tom-demeyer#list.

18. See Eduardo Reck Miranda and Marcelo M. Wanderley, *New Digital Musical Instruments: Control and Interaction beyond the Keyboard*, vol. 21 (Middleton, WI: A-R Editions, 2006).

19. See Joseph Paradiso, "Current Trends in Electronic Music Interfaces: Guest Editors Introduction," *Journal of New Music Research* 32, no. 4 (March 1988): 345–349.

20. Robert L. Adams, Michael Brook, John Eichenseer, Mark Goldstein, and Geoff Smith, Electronic Musical Instrument, US Patent 6,005,181, filed April 7, 1998, and issued December 21, 1999.

21. Ryan "Effort and Expression," 415.

22. Ryan, "Effort and Expression," 416.

23. There is a difference in the amount of acceleration that a moving human body can produce in comparison with a machine. When we make an acceleration, we can

produce a maximum of 2 g, whereas an industrial washing machine spinning produces a g-force of almost 600 g!

Chapter 3

1. See, for instance, Thomas Fysh and J. F. Thompson, "A Wii Problem," *Journal of the Royal Society of Medicine* 102, no. 12 (December 2009): 502; and Maarten B. Jalink, Erik Heineman, Jean-Pierre E. N. Pierie, and Henk O. ten Cate Hoedemaker, "Nintendo Related Injuries and Other Problems: Review," *BMJ* 349, December 16, 2014, https://www.bmj.com/content/349/bmj.g7267.

2. Thomas Ricker, "Nintendo Addresses Wiimote Damage," *Engadget*, December 6, 2006, https://www.engadget.com/2006-12-06-nintendo-addresses-wiimote-damage-issues-sends-email.html?]&guccounter=1.

3. ADXL 330 Data Sheet, Analog Devices, accessed December 7, 2020, https://www.analog.com/en/products/adxl330.html#product-overview.

4. "Dedicated Video Game Sales Units," Nintendo, accessed December 7, 2020, https://www.nintendo.co.jp/ir/en/finance/hard_soft/.

5. Seth Schiesel, "Motion, Sensitive," *New York Times*, November 16, 2010, https://www.nytimes.com/2010/11/28/arts/video-games/28video.html.

6. For more information on the relationship between game controllers and physical experience, see David Parisi, "Game Interfaces as Bodily Techniques," in *Gaming and Simulations: Concepts, Methodologies, Tools and Applications*, vol. 1, ed. Mehdi Khosrow-Pour (Hershey, PA: IGI Global, 2011), 1033–1047.

7. Asaf Gurner, Optical Instrument with Tone Signal Generating Means, U.S. Patent 5,045,687, filed May 10, 1989, and issued September 3, 1991.

8. Asaf Gurner, "Light Harp at CES 1993," presentation at CES 1993, video, 2:30, posted November 25, 2007, https://www.youtube.com/watch?v=YoxsnCiX05k&feature=emb_title&ab_channel=AssafGurner.

9. Alex Dunn, "Sega Activator Training Video," video, 4:09, posted August 25, 2006, https://www.youtube.com/watch?v=ql-UZv3AS-E.

10. Dunn, "Sega Activator Training Video."

11. Screenshot in Eric Frederiksen, "Eric's Biggest Tech Regret: The Sega Activator," https://www.technobuffalo.com/erics-biggest-tech-regret-the-sega-activator, accessed May 3, 2021; capitalized in original.

12. See http://www.jaronlanier.com/.

13. Antonin Artaud, *The Theater and Its Double*, trans. M. C. Richards (New York: Grove Press, 1965), 49.

14. Wayne Carlson, *Computer Graphics and Computer Animation: A Retrospective Overview* (Columbus: Ohio State University Press, 2017), 525.

15. "Military," Polhemus, accessed April 10, 2020, https://polhemus.com/applications/military-old.

16. Mark Weiser, Rich Gold, and John Seely Brown, "The Origins of Ubiquitous Computing Research at PARC in the Late 1980s," *IBM Systems Journal* 38, no. 4 (1999): 693–696.

17. Mark Weiser, "The Computer for the 21st Century," *Scientific American* 265, no. 3 (1991): 94–105.

18. Sensorimotor abilities in disabled can still be expressive, even if limited. See Simon Penny, "Sensorimotor Debilities in Digital Cultures," *AI & SOCIETY*, 2021, https://doi.org/10.1007/s00146-021-01186-0.

19. Nathan Chandler, "How the Nintendo Power Glove Worked," HowStuffWorks, March 25, 2015, https://electronics.howstuffworks.com/nintendo-power-glove.htm.

20. Dana L. Gardner, "Inside Story On: The Power Glove," *Design News* 45, no. 23 (December 4, 1989): 63–72, https://www.microsoft.com/buxtoncollection/a/pdf/PowerGlove%20Design%20News%20%20Article.pdf.

21. Sangbeom Kim, Ian Lamont, Hiroshi Ogasawara, Mansoo Park, and Hiroaki Takaoka, *Nintendo's Revolution* (MIT Sloan Management School Report, October 18, 2011), https://www.yumpu.com/en/document/read/10783464/nintendos-revolution-mit-sloan-school-of-management.

22. For more on path dependency, see Paul A. David, "Clio and the Economics of QWERTY," *American Economic Review* 75, no. 2 (1985): 332–337.

23. Adam Champy, "Elements of Motion: 3D Sensors in Intuitive Game Design," *Analog Dialogue*, April 2007, https://www.analog.com/ru/analog-dialogue/articles/3d-sensors-in-intuitive-game-design.html.

24. For a critique of natural interfaces, see Donald Norman, "Natural User Interfaces Are Not Natural," *Interactions* 17, no. 3 (2010): 6–10, https://interactions.acm.org/archive/view/may-june-2010/natural-user-interfaces-are-not-natural1.

25. There have been several versions of the Kinect, and they have used different types of technology. I have focused here on the first version, which used the structured light approach.

26. A similar camera-based recognition system was released by Sony in 2006. Sony's Move controller was designed to work with the PlayStation 3 console. The system

Notes

utilized Sony's Eye camera and could track the handheld Move controller, which was embedded with colored LEDs, as well as a gyroscope, accelerometer, and magnetic sensor to track the controller in three dimensions.

27. A complementary metal-oxide-semiconductor sensor has some special characteristics versus the earlier charge-coupled device sensors found in most digital cameras. Besides being easier and cheaper to manufacture and having lower power consumption, the sensor also enables software to access every single pixel individually and to perform fast processing at each pixel point.

28. For more technical detail, see Hamed Sarbolandi, Damien Lefloch, and Andreas Kolb, "Kinect Range Sensing: Structured-Light versus Time-of-Flight Kinect," *Computer Vision and Image Understanding* 139 (2015): 1–20.

29. The initial story surfaced in PC magazine and Game Stop in the UK. See https://www.npr.org/sections/thetwo-way/2010/11/05/131092329/xbox-kinect-not-racist-after-all, accessed May 30, 2021.

30. The Wii system also had an infrared-based sensor installed in an accompanying sensor bar that, with help of an IR emitter in the Wiimote, could locate the controller in space.

31. A *pulse oximeter* is a noninvasive photoplethysmogram (PPG) sensor that measures the amount of oxygen in the blood by detecting how much light gets absorbed by the blood with each pulse.

32. Kaprow's initial phrasing spoke of the difference between art-like art and like-like art. See Allan Kaprow, "The Real Experiment," in *Essays on the Blurring of Art and Life*, expanded edition, ed. Jeff Kelley (Berkeley: University of California Press, 2003), 201–218.

Chapter 4

1. See https://borderless.teamlab.art/.

2. Naomi Rea, "teamLab's Tokyo Museum Has Become the World's Most Popular Single-Artist Destination, Surpassing the Van Gogh Museum," Artnet News, August 7, 2019, https://news.artnet.com/exhibitions/teamlab-museum-attendance-1618834.

3. Shuhei Senda, "Mori Building and teamLab to Launch Mori Building Digital Art Museum in Toyko This Summer," designboom, May 1, 2018, https://www.designboom.com/art/mori-building-teamlab-digital-art-museum-tokyo-05-01-2018.

4. For more details on computer vision techniques, see a standard textbook, such as Richard Sziliski, *Computer Vision: Algorithms and Applications* (London: Springer, 2011). Or, for a more arts-based introduction, see The Coding Train, Computer Vision, YouTube playlist of videos presented by Daniel Shiffman, last updated July

21, 2016, https://www.youtube.com/playlist?list=PLRqwX-V7Uu6aG2RJHErXKSWF DXU4qo_ro.

5. Marshall McLuhan, "The Medium Is the Message," in *Understanding Media: The Extensions of Man* (Cambridge, MA: MIT Press, 1994), 8–9.

6. See Allan Kaprow, "Happenings in the New York Scene," in Kaprow, *Essays on the Blurring of Art and Life*, expanded edition, ed. Jeff Kelley (Berkeley: University of California Press, 2003), 15–26.

7. Kaprow, cited in Julie H. Reiss, *From Margin to Centre: The Spaces of Installation Art* (Cambridge, MA: MIT Press, 1999), 24.

8. Roy Ascott, quoted by Brian Eno, *A Year with Swollen Appendices: Brian Eno's Diary* (London: Faber & Faber, 1996), 368.

9. See Meow Wolf, http://www.meowwolf.com; Dreamscape Immersive, http://www.dreamscapeimmersive.com; Moment Factory, http://www.momentfactory.com; Marshmallow Laser Feast, http://www.marshmallowlaserfeast.com; Phenomena, https://www.thephenomenavr.com; Superblue, http://www.superblue.com; and Punchdrunk, http://www.punchdrunk.com.

10. Joseph B. Pine and James H. Gilmore, "Welcome to the Experience Economy," *Harvard Business Review* 76 (1998): 97–105.

11. See http://dreamscapeimmersive.com.

12. Rachel Monroe, "Can an Art Collective Become the Disney of the Experience Economy?," *New York Times Magazine*, May 1, 2019, https://www.nytimes.com/interactive/2019/05/01/magazine/meow-wolf-art-experience-economy.html.

13. For more detail on these immersive histories from an art historical perspective, see Oliver Grau, *Virtual Art: From Illusion to Immersion* (Cambridge, MA: MIT Press, 1996); and Erkki Huhtamo, *Illusions in Motion: Media Archaeology of the Moving Panorama and Related Spectacles* (Cambridge, MA: MIT Press, 2013).

14. See Arnold Aronson, *The History and Theory of Environmental Scenography* (London: Bloomsbury Publishing, 2018), 117. See also Elena Filipovic, "A Museum That Is Not," e-flux, no. 4 (March 2009), https://www.e-flux.com/journal/04/68554/a-museum-that-is-not/#_ftn32.

15. László Moholy-Nagy, "Theater, Circus, Variety," in *The Theater of the Bauhaus*, ed. Walter Gropius and Arthur S. Wensinger (London: Metheun, 1979). Or in the original German: László Moholy-Nagy, Oskar Schlemmer, and Farkas Molnar, *Die Bühne im Bauhaus* (Mainz: Florian Kupferberg, 1965).

16. Robert Hughes, "Paradise Now," *Guardian*, March 20, 2006, https://www.theguardian.com/artanddesign/2006/mar/20/architecture.modernism1.

17. Lyubov Popova, untitled manuscript, in *Women Artists of the Russian Avant-Garde 1910–1930*, ed. Krystyna Gmurzynska (Cologne: Galerie Gmurzynska 1980), 68.

18. Artaud, *Theater and Its Double*, 95–96.

19. Walter Benjamin, "The Artwork in the Age of Mechanical Reproduction," in *Illuminations*, trans. Harry Zohn (New York: Schocken, 2007), 217–252.

20. Benjamin, "Artwork," 237.

21. Years later, one of Benjamin's most astute commentators, American European intellectual historian Susan Buck-Morss, put it even more forcefully. The technoaesthetics that Benjamin describes of cinema and the consumer spectacles of modern Paris at the turn of the century both expose the senses to physical shock, which corresponds to psychic shock. The goal is thus to protect the senses from the flood of stimuli by deadening them—to numb the organism in order to repress the trauma of modern life. See Susan Buck-Morss, "Aesthetics and Anaesthetics: Walter Benjamin's Artwork Essay Reconsidered," *October* 62 (1992): 3–41.

22. Siegfried Kracauer, *From Caligari to Hitler: A Psychological History of the German Film* (Princeton, NJ: Princeton University Press, 2019), 301.

23. Benjamin, "Artwork," 251.

24. Benjamin, "Artwork," 241–242.

25. American media historian Fred Turner put it this way: "Immersive, multimediated environments designed to expand individual consciousness and a sense of membership in the human collective first came into being as part of the same urge to defeat the forces of totalitarianism that animated the most aggressive cold warriors." Fred Turner, *The Democratic Surround: Multimedia and American Liberalism from World War II to the Psychedelic Sixties* (Chicago: University of Chicago Press, 2013), 8–9. For a discussion of the collaboration between artists and engineers within the backdrop of Cold War military-industrial research, see W. Patrick McCray, *Making Art Work: How Cold War Engineers and Artists Forged a New Creative Culture* (Cambridge, MA: MIT Press, 2020); and John Beck and Ryan Bishop, *Technocrats of the Imagination: Art, Technology, and the Military-Industrial Avant-Garde* (Durham, NC: Duke University Press, 2020).

26. The term *emancipated spectator* derives from the work of French cultural critic Jacques Rancière: "Emancipation starts from the principle of equality. It begins when we dismiss the opposition between looking and acting and understand that the distribution of the visible itself is part of the configuration of domination and subjection. It starts when we realize that looking is also an action that confirms or modifies that distribution, and that 'interpreting the world' is already a means of

transforming it, of reconfiguring it." See Jacques Rancière, *The Emancipated Spectator*, trans. Gregory Elliot (London: Verso, 2009).

27. See Nam June Paik, "Cybernated Art," in *The New Media Reader*, ed. Noah Waldrip-Fruin and Nick Montfort (Cambridge, MA: MIT Press, 2003), 227–228; Umberto Eco and Bruno Munari, *Arte programmata e cinetica, 1953–1963: l'ultima avanguardia* (Milan: G. Mazzotta, 1983); and Jack Burnham, "Systems Esthetics," *Artforum* 7, no. 1 (1968): 30–35.

28. See Branden W. Joseph, *Random Order: Robert Rauschenberg and the Neo-Avant-Garde* (Cambridge, MA: MIT Press, 2003), 245–249.

29. Maurice Tuchman, *Art & Technology: A Report on the Art & Technology Program of the Los Angeles County Museum of Art, 1967–1971* (Los Angeles: LACMA, 1971), 279–288.

30. "Teledyne Technologies Inc. History," FundingUniverse, accessed December 21, 2020, http://www.fundinguniverse.com/company-histories/teledyne-technologies-inc-history/.

31. Tuchman, *Art & Technology*, 9–14.

32. Max Kozloff, "The Multimillion Dollar Art Boondoggle," *Artforum* 10, no. 2 (1971): 74.

33. For extensive documentation, see Vincent Bonin, "9 Evenings: Theatre and Engineering Fonds," Daniel Langlois Foundation, 2006, https://www.fondation-langlois.org/html/e/page.php?NumPage=294.

34. "Variations VII," John Cage Trust, accessed June 29, 2019, https://johncage.org/pp/John-Cage-Work-Detail.cfm?work_ID=272.

35. Branden W. Joseph, ed., *October Files #4: Robert Rauschenberg* (Cambridge, MA: MIT Press, 2002), 24.

36. Tuchman, *Art & Technology*, 127–142.

37. Tuchman, *Art & Technology*, 140.

38. Rebecca Lemov, "Running Amok in Labyrinthine Systems: The Cyber-Behaviorist Origins of Soft Torture," *Limn*, no. 1: Systemic Risk, ed. Stephen J. Collier, Christopher M. Kelty, and Andrew Lakoff (2011), https://limn.it/articles/running-amok-in-labyrinthine-systems-the-cyber-behaviorist-origins-of-soft-torture/.

39. For an extensive discussion of Pask, see Andrew Pickering, *The Cybernetic Brain: Sketches for Another Future* (Chicago: University of Chicago Press, 2010), 309–378.

40. Nobert Wiener, *Cybernetics: Or Control and Communication in the Animal and the Machine* (Cambridge, MA: MIT Press, 2007), 11.

41. For more about social and cultural cybernetics projects, see Pickering, *Cybernetic Brain*; Steve J. Heims, *Constructing a Social Science for Postwar America: The Cybernetics Group, 1946–1953* (Cambridge, MA: MIT Press, 1993); and Orit Halpern, *Beautiful Data* (Durham, NC: Duke University Press, 2014).

42. Cedric Price and Joan Littlewood, "The Fun Palace," *TDR: The Drama Review* 12, no. 3 (1968): 130.

43. Stanley Matthews, "The Fun Palace: Cedric Price's Experiment in Architecture and Technology," *Technoetic Arts: A Journal of Speculative Research* 3, no. 2 (2005): 80.

44. It's not clear from the existing notes what exactly these "electronic sensors" would have been.

45. Price and Littlewood, "Fun Palace," 134.

46. Pickering, *Cybernetic Brain*, 371.

Chapter 5

1. Itsuki Doi, Takashi Ikegami, Atsushi Masumori, Hiroki Kojima, Kohei Ogawa, and Hiroshi Ishiguro, "A New Design Principle for an Autonomous Robot," in *Proceedings of the 14th European Conference on Artificial Life ECAL 2017* (Cambridge, MA: MIT Press, 2017), 490–496.

2. Donna Haraway, "A Cyborg Manifesto: Science, Technology and Socialist Feminism in the Late Twentieth Century," in *Simians, Cyborgs, and Women: The Reinvention of Nature* (London: Routledge, 1990), 151–152.

3. See the Ishiguro Lab website, https://eng.irl.sys.es.osaka-u.ac.jp/; and the Ikegami Lab website, https://www.sacral.c.u-tokyo.ac.jp/.

4. Doi et al., "New Design Principle," 490.

5. See W. R. Ashby, "Principles of the Self-Organizing Dynamic System," *Journal of General Psychology* 37, no. 2 (1947): 125–128.

6. See Yuto Miyamoto, "Inside Alternative Machine," *Evertale* (blog), November 3, 2018, https://medium.com/evertale-english/alternative-machine-7058e71be53. For classic A-Life sources, see *Artificial Life: An Overview*, ed. Christopher G. Langton (Cambridge, MA: MIT Press, 1997). For an anthropological account, see Stefan Helmreich, *Silicon Second Nature: Culturing Artificial Life in a Digital World* (Berkeley: University of California Press, 1998). And for a more popular rendering, see Kevin Kelly, *Out of Control: The Rise of Neo-Biological Civilization* (Reading, MA: Addison-Wesley Longman, 1994).

7. For another anthropological take on this question, see Jannik Friberg Lindegaard and Lars Rune Christensen, "Allusive Machines: Encounters with Android Life," in

NordCHI: Proceedings of the 10th Nordic Conference on Human-Computer Interaction (New York: Association for Computing Machinery, 2018), 114–124.

8. See *AI: More Than Human*, exhibition catalog (London: Barbican Centre, 2019).

9. See Tom Froese, Nathaniel Virgo, and Eduardo Izquierdo, "Autonomy: A Review and a Reappraisal," *Advances in Artificial Life* (Berlin: Springer, 2007), 455–464.

10. For an accessible introduction to autopoiesis, see Humberto R. Maturana and Francisco J. Varela, *The Tree of Knowledge: The Biological Roots of Human Understanding* (Boston: New Science Library/Shambhala Publications, 1987).

11. Takashi Ikegami in discussion with the author, June 2020. All subsequent quotes, unless otherwise noted, are from the same session.

12. Takashi Ikegami in discussion with the author, June 2020.

13. See Hiroshi Ishiguro Laboratories, accessed December 5, 2020, http://www.geminoid.jp/en/index.html. See also Alastair Gale and Takashi Mochizuki, "Robot Hotel Loses Love for Robots," *Wall Street Journal*, January 14, 2019, https://www.wsj.com/articles/robot-hotel-loses-love-for-robots-11547484628.

14. See Norihiro Maruyama, Mizuki Oka, and Takashi Ikegami, "Creating Space-Time Affordances via an Autonomous Sensor Network," *2013 IEEE Symposium on Artificial Life* (New York: IEEE, 2013): 67–73.

15. Maruyama et al., "Creating Space-Time Affordances," 68.

16. See Alan Mathison Turing, "The Chemical Basis of Morphogenesis," *Bulletin of Mathematical Biology* 52, no. 1–2 (1990): 153–197.

17. Doi et al., "New Design Principle," 493.

18. See Cathy O'Neil, *Weapons of Math Destruction* (New York: Crown, 2017); Meredith Broussard, *Artificial Unintelligence: How Computers Mistake the World* (Cambridge, MA: MIT Press, 2018); and Safiya Umoja Noble, *Algorithms of Oppression* (New York: NYU Press, 2018).

19. Warren S. McCulloch and Walter Pitts, "A Logical Calculus of the Ideas Immanent in Nervous Activity," *Bulletin of Mathematical Biophysics* 5, no. 4 (1943): 115–133.

20. Manfred Spitzer, *The Mind within the Net: Models of Learning, Thinking, and Acting* (Cambridge, MA: MIT Press, 1999), 5.

21. McCulloch and Pitts, "Logical Calculus," 116.

22. Friedrich August Hayek, *The Sensory Order: An Inquiry into the Foundations of Theoretical Psychology* (Chicago: University of Chicago Press, 1952). Although the book is mainly a psychology treatise, Hayek is better known for his development and

promotion of the philosophical and economic free-market ideology known as *neoliberalism*. See Philip Mirowski, *Never Let a Serious Crisis Go to Waste: How Neoliberalism Survived the Financial Meltdown* (New York: Verso, 2013); and Quinn Slobodian, *Globalists: The End of Empire and the Birth of Neoliberalism* (Cambridge, MA: Harvard University Press, 2020).

23. Friedrich August Hayek, *Hayek on Hayek: An Autobiographical Dialogue* (Chicago: University of Chicago Press, 1994), 122.

24. Hayek, *Sensory Order*, 39.

25. Hayek, *Sensory Order*, 142.

26. Hayek, *Sensory Order*, 52.

27. Hayek, *Sensory Order*, 167.

28. While *The Sensory Order* never alludes to economics, commentators have made clear that Hayek's model of spontaneous order that exemplifies the mind can be generalized into other phenomena—namely, the operation of markets.

29. Friedrich August Hayek, "Theory of Complex Phenomena," in Hayek, *Studies in Philosophy, Politics and Economics* (Chicago: University of Chicago Press, 1980), 22–42.

30. D. E. Rumelhart and J. C. McClelland, eds., *Parallel Distributed Processing*, vol. 1 (Cambridge, MA: MIT Press, 1999), ix.

31. Francisco Varela, Eleanor Rosch, and Evan Thompson, *The Embodied Mind: Cognitive Science and Human Experience* (Cambridge, MA: MIT Press, 1991), 368.

32. "Hayek made a quite fruitful suggestion, made contemporaneous by the psychologist Donald Hebb [about] the reinforcement of the connection between neural cells. These days this is known as a Hebbian synapse but Hayek quite independently came upon the idea. I think the essence of his analysis still remains with us." See Gerald Edelman, *Neural Darwinism: The Theory of Neuronal Group Selection* (New York: Basic Books, 1987).

33. Eugene M. Izhikevich, "Simple Model of Spiking Neurons," *IEEE Transactions on Neural Networks* 14, no. 6 (2003): 1569–1572.

34. Takashi Ikegami in discussion with the author, June 2020.

35. Takashi Ikegami in discussion with the author, March 2019.

36. See Michael Allcock, dir., *Fear of Dancing: An Offbeat Film about Dance* (Toronto, Canada: Tortuga Films, 2020).

37. The term *affordance* comes from the work of the American psychologist James J. Gibson, who defines it as "The affordances of the environment are what it offers

the animal, what it provides or furnishes, either for good or ill." See James J. Gibson, *The Ecological Approach to Visual Perception* (New York: Psychology Press, 2014), 119.

Chapter 6

1. Most of the reports from Dickmanns's earlier research can be found in his extensive online archives at http://dyna-vision.de/. See also Ernst Dickmanns, *Dynamic Vision for Perception and Control of Motion* (London: Springer, 2007), 37–38.

2. Dynamic vision used a process called *prediction error feedback*, in which feedback from the real world would error-correct false predictions made by the computer model. This technique took its form from a complex discipline called *control engineering*. As one of the bases for Norbert Wiener's vision of cybernetics, control engineering tries to direct machines to reach programmed goals by using feedback to solve problems, like how to direct a machine to reach a goal using its own feedback. Then you can make a prediction about how this should develop over time. Dickmanns's system thus tried to predict the state in which an object might be as time changed based on the feedback the system received from the world. See Dickmanns, *Dynamic Vision* for more detail.

3. Sensor fusion is used in any situation with multiple sensors—for example, self-driving cars. One widely used technique called a *Kalman filter* aims to understand the state of a system as it evolves in time. Based on its past and its present, it attempts to make predictions about the future state of a system. The Kalman filter is an application of the more general concepts of Markov chains and Bayesian inference, which are mathematical systems that iteratively refine their guesses using evidence. These are tools designed to help science itself test ideas and are the basis of what we call *statistical significance*.

4. See Vassilis Cutsuridis, "Cognitive Models of the Perception-Action Cycle: A View from the Brain," *The 2013 International Joint Conference on Neural Networks (IJCNN)* (New York: IEEE, 2013), 1–8. See also Joaquín M. Fuster, "Physiology of Executive Functions: The Perception-Action Cycle," in *Principles of Frontal Lobe Function*, ed. Donald T. Stuss and Robert T. Knight (Oxford: Oxford University Press, 2002), 96–108.

5. The DARPA Grand Challenge is an annual prize-driven competition organized by the Defense Advanced Research Projects Agency (DARPA) for autonomous vehicles. See "The Grand Challenge," DARPA, accessed December 9, 2020, https://www.darpa.mil/about-us/timeline/-grand-challenge-for-autonomous-vehicles.

6. Despite this engineering sophistication, self-driving cars are by no means foolproof. For a strong critique, see Broussard, *Artificial Unintelligence*, 121–148.

7. See Peter Kleinschmidt, "How Many Sensors Does a Car Need?," *Sensors and Actuators* 31, no. 1–3 (March 1992): 35–45.

Notes

8. Neil Tyler, "Demand for Automotive Sensors Is Booming," *New Electronics*, December 14, 2016, https://bit.ly/3gwV8zf.

9. Silvia Casini, "Synesthesia, Transformation and Synthesis: Toward a Multisensory Pedagogy of the Image," *Senses and Society* 12, no. 1 (2017): 5.

10. Andrew Gross, "Think You're in Your Car More? You're Right. Americans Spend 70 Billion Hours behind the Wheel," AAA Newsroom, February 27, 2019, https://newsroom.aaa.com/2019/02/think-youre-in-your-car-more-youre-right-americans-spend-70-billion-hours-behind-the-wheel/.

11. "Road Traffic Injuries," World Health Organization, February 7, 2020, https://www.who.int/news-room/fact-sheets/detail/road-traffic-injuries.

12. Kristen Hall-Geisler, "How Anti-sleep Alarms Work," HowStuffWorks, February 4, 2009, https://electronics.howstuffworks.com/gadgets/automotive/anti-sleep-alarm.htm. See also Anne Eisenberg, "What's Next: A Passenger Whose Chatter Is Always Appreciated," *New York Times*, December 27, 2001, https://www.nytime.com/2001/12/27/technology/what-s-next-a-passenger-whose-chatter-is-always-appreciated.html.

13. T. Maeda, H. Ando, T. Amemiya, N. Nagaya, M. Sugimoto, and M. Inami, "Shaking the World: Galvanic Vestibular Stimulation as a Novel Sensation Interface," in *ACM SIGGRAPH 2005 Emerging Technologies*, ed. Donna Cox (Los Angeles: ACM, 2005), 17.

14. Sadayuki Tsugawa, "Trends and Issues in Safe Driver Assistance Systems: Driver Acceptance and Assistance for Elderly Drivers," *IATSS Research* 30, no. 2 (2006): 6–18.

15. Discussion with the head strategic engineer at the Leipzig BMW factory, July 2017.

16. Marco Allegretti and Silvano Bertoldo, "Cars as a Diffuse Network of Road-Environment Monitoring Nodes," *Wireless Sensor Network* 6, no. 9 (2014): 184–191.

17. Kara Swisher, "Amazon Isn't Interested in Making the World a Better Place," *New York Times*, February 15, 2019, https://www.nytimes.com/2019/02/15/opinion/amazon-new-york-hq2.html.

18. John Markoff, "Can't Find a Parking Spot? Check Smartphone," *New York Times*, July 12, 2008, https://www.nytimes.com/2008/07/12/business/12newpark.html.

19. Dutch sociologist of science Karin Bijsterveld uses the terms *acoustic* and *techno cocooning* to describe how the conditioning of sound achieved by advanced technologies has been specifically engineered to create the sense of individual control and privacy within particular spaces. Bijsterveld points to the development of car radio systems starting in the late 1920s in Europe as a strong example of how technologies were adapted to changing social and cultural conditions of mobility. The main shift

over time is that drivers were given technologies that allowed them to gain increasing control over the acoustic design of the interior of their vehicles. The sensory acoustic experience of driving was thus radically transformed. It moved from "what the consumer does with the product" to a post-1970 paradigm in which auto manufacturers became preoccupied with "what the product does to the consumer." See Bijsterveld, "Acoustic Cocooning: How the Car Became a Place to Unwind," *Senses and Society* 5, no. 2 (2010): 189–211.

20. Alex Sobran, "Revisting the Original Uber Audio Upgrade: The Blaupunkt Berlin," Petrolicious, June 14, 2017, https://petrolicious.com/articles/revisiting-the-original-uber-audio-upgrade-the-blaupunkt-berlin.

21. In German, the term is *Störgeräuschabhängige Lautstärkesteuerung* (SALS).

22. For an introduction to psychoacoustics, see Perry R. Cook, ed., *Music, Cognition, and Computerized Sound: An Introduction to Psychoacoustics* (Cambridge, MA: MIT Press, 2001).

23. The microphones are a special type called *binaural*, which focus on recording what the ear hears.

24. "Bose Introduces QuietComfort Road Noise Control," Bose, January 8, 2019, https://globalpressroom.bose.com/us-en/pressrelease/view/1966.

25. "Sound Symposer Explained," TeamSpeed, August 13, 2012, https://teamspeed.com/forums/991-997-996/75154-991-sound-symposer-explained.html.

26. Kenji Mori, Naoto Kitagawa, Akihiro Inukai, and Simon Humphries, Vehicle Expression Operation Control System, Vehicle Communication System, and Vehicle which Performs Expression Operation, US Patent 6757593B2, filed March 18, 2002, and issued June 29, 2004.

27. Mori et al., Vehicle Expression Operation Control System, 25.

28. Mori et al., Vehicle Expression Operation Control System, 24.

29. "The Car that Responds to Your Mood: New Jaguar and Rover Tech Helps Reduce Stress," Jaguar Land Rover, July 9, 2019, https://www.jaguarlandrover.com/news/2019/07/car-responds-your-mood-new-jaguar-land-rover-tech-helps-reduce-stress.

30. See Shira Ovide, "The Case for Banning Facial Recognition," *New York Times*, June 9, 2020, https://www.nytimes.com/2020/06/09/technology/facial-recognition-software.html; and http://gendershades.org/.

31. Mike Elgan, "What Happens When Cars Get Emotional?," *Fast Company*, June 27, 2019, https://www.fastcompany.com/90368804/emotion-sensing-cars-promise-to-make-our-roads-much-safer.

32. "Natürliche und vollständig multimodale Interaktion mit dem Fahrzeug und der Umgebung. Auf dem Mobile World Congress 2019 präsentiert die BMW Group erstmals BMW Natural Interaction," BMW Group, February 25, 2019, https://tinyurl.com/unm89r85.

33. See https://www.moodify.today/.

34. Jack Stuster, *Aggressive Driving Enforcement: Evaluations of Two Demonstration Programs* (Washington, DC: US Department of Transportation/National Highway Traffic Safety Administration, March 2004), https://one.nhtsa.gov/people/injury/research/aggdrivingenf/.

35. Hans Selye, "The Nature of Stress," *Basal Facts* 7, no. 1 (1985): 3–11.

36. Pablo E. Paredes, Francisco Ordoñez, Wendy Ju, and James A. Landay, "Fast & Furious: Detecting Stress with a Car Steering Wheel," in *Proceedings of the 2018 CHI Conference on Human Factors in Computing Systems* (Montreal: ACM Press, 2018), 1–12.

37. Wan-Young Chung, Teak-Wei Chong, and Boon-Giin Lee, "Methods to Detect and Reduce Driver Stress: A Review," *International Journal of Automotive Technology* 20, no. 5 (2019): 1051–1063.

38. Mariam Hassib, Michael Braun, Bastian Pfleging, and Florian Alt, "Detecting and Influencing Driver Emotions Using Psycho-physiological Sensors and Ambient Light," in *Human-Computer Interaction—INTERACT 2019* (Cham, Switzerland: Springer, 2019), 721–742.

39. See "The World of Air Transport in 2019," ICAO Report, 2019, https://www.icao.int/annual-report-2019/Pages/the-world-of-air-transport-in-2019.aspx#:~:text=According%20to%20ICAO's%20preliminary%20compilation,a%201.7%20per%20cent%20increase.

40. Pablo E. Paredes, Yijun Zhou, Nur Al-Huda Hamdan, Stephanie Balters, Elizabeth Murnane, Wendy Ju, and James. A. Landay, "Just Breathe: In-Car Interventions for Guided Slow Breathing," in *Proceedings of the ACM on Interactive, Mobile, Wearable and Ubiquitous Technology* 2, no. 1 (March 2018): 1–23.

Chapter 7

1. "Finding a life path as sweet as nectar in "measuring taste," Kyushu University, accessed December 10, 2020, https://www.kyushu-u.ac.jp/en/university/professor/toko.html.

2. K. Hayashi, M. Yamanaka, K. Toko, and K. Yamafuji, "Multichannel Taste Sensor Using Lipid Membranes," *Sensors and Actuators B: Chemical* 2, no. 3 (1990): 205–213.

3. For a survey of e-tongue technology, see Y. Tahara and K. Toko, "Electronic Tongues–A Review," *IEEE Sensors Journal* 13, no. 8 (August 2013): 3001–3011.

4. See https://elbarri.com/en/restaurant/tickets/.

5. John McQuaid, *Tasty: The Art and Science of What We Eat* (New York: Scribner, 2015), 17.

6. Tammy La Gorce, "The Tastemakers," *New Jersey Monthly*, January 17, 2011, https://njmonthly.com/articles/eat-drink/the-tastemakers/.

7. See Joel Fuhrman, "The Hidden Dangers of Fast and Processed Food," *American Journal of Lifestyle Medicine* 12, no. 5 (2018): 375–381; and Leonie Elizabeth, Priscila Machado, Marit Zinöcker, Phillip Baker, and Mark Lawrence, "Ultra-Processed Foods and Health Outcomes: A Narrative Review," *Nutrients* 12, no. 7 (2020): 1955.

8. "Mouthfeel & Texturisation: A Dive into the Food & Beverage Industry," Asia Pacific Food Industry, December 5, 2018, https://apfoodonline.com/industry/mouthfeel-and-texturisation-a-dive-into-the-food-beverage-industry/.

9. See David Julian McClements, *Future Foods: How Modern Science Is Transforming the Way We Eat* (Stuttgart: Springer Nature, 2019), 85

10. Harry T. Lawless and Hildegarde Heymann, *Sensory Evaluation of Food: Principles and Practices* (New York: Springer, 2010), 1.

11. See https://monell.org.

12. Lawless and Heymann, *Sensory Evaluation*, 2.

13. "Sensory Science," Cargill, accessed June 20, 2020, https://www.cargill.com/about/research/sensory-science.

14. Christy Spackman and Jacob Lahne, "Sensory Labor: Considering the Work of Taste in the Food System," *Food, Culture & Society* 22, no. 2 (2019): 142–151.

15. Lawless and Heymann, *Sensory Evaluation*, 1.

16. David Howes, "The Science of Sensory Evaluation: An Ethnographic Critique," in *The Social Life of Materials*, ed. Adam Drazin and Susanne Küchler (London: Bloomsbury, 2015), 81–97.

17. Da-Wen Sun, "Inspecting Pizza Topping Percentage and Distribution by a Computer Vision Method," *Journal of Food Engineering* 44, no. 4 (2000): 245–249.

18. Claudia Gonzalez Viejo, Damir D. Torrico, Frank R. Dunshea, and Sigfredo Fuentes, "Bubbles, Foam Formation, Stability and Consumer Perception of Carbonated Drinks: A Review of Current, New and Emerging Technologies for Rapid Assessment and Control," *Foods* 8, no. 12 (2019): 596.

19. M. S. Thakur and K. V. Ragavan, "Biosensors in Food Processing," *Journal of Food Science and Technology* 50, no. 4 (2013): 625–641.

20. Kate Murphy, "Food Processors Rely on Electronic Sensors," *New York Times*, November 3, 1997, https://archive.nytimes.com/www.nytimes.com/library/cyber/week/110397food.html.

21. Murphy, "Food Processors Rely on Electronic Sensors."

22. Maria L. Rodríguez-Méndez, José A. De Saja, Rocio González-Antón, Celia García-Hernández, Cristina Medina-Plaza, Cristina García-Cabezón, and Fernando Martín-Pedrosa, "Electronic Noses and Tongues in Wine Industry," *Frontiers in Bioengineering and Biotechnology* 4 (2016): 81.

23. P. Hauptmann, R. Borngraeber, J. Schroeder, and J. Auge, "Artificial Electronic Tongue in Comparison to the Electronic Nose. State of the Art and Trends," in *Proceedings of the 2000 IEEE/EIA International Frequency Control Symposium and Exhibition* (Kansas City, MO: IEEE, 2000), 22–29.

24. Wenwen Hu, Liangtian Wan, Yingying Jian, Cong Ren, Ke Jin, Xinghua Su, Xiaoxia Bai, Hossam Haick, Mingshui Yao, and Weiwei Wu, "Electronic Noses: From Advanced Materials to Sensors Aided with Data Processing," *Advanced Materials Technologies* 4, no. 2 (2019): 1800488.

25. Paolo Pelosi, Jiao Zhu, and Wolfgang Knoll, "From Gas Sensors to Biomimetic Artificial Noses," *Chemosensors* 6, no. 32 (2018): 32.

26. Michael J. Schöning, Peter Schroth, and Stefan Schütz, "The Use of Insect Chemoreceptors for the Assembly of Biosensors Based on Semiconductor Field-Effect Transistors," *Electroanalysis: An International Journal Devoted to Fundamental and Practical Aspects of Electroanalysis* 12, no. 9 (2000): 645–652.

27. Hervé This, *Molecular Gastronomy: Exploring the Science of Flavor* (New York: Columbia University Press, 2006), 334.

28. The classic definition of *gastronomy* comes from Brillat-Savarin. See Jean Brillat-Savarin, *The Physiology of Taste: or Meditations on Transcendental Gastronomy*, trans. M. K. Fisher (New York: Vintage, 2011).

29. "Estimating the Burden of Foodborne Diseases," World Health Organization, accessed December 10, 2020, https://www.who.int/activities/estimating-the-burden-of-foodborne-diseases.

30. "Scottish Engineers Develop Artificial 'Tongue' with a Taste for Spotting Fake Whisky," CBC Radio, August 6, 2019, https://www.cbc.ca/radio/asithappens/as-it-happens-tuesday-edition-1.5237385/scottish-engineers-develop-artificial-tongue-with-a-taste-for-spotting-fake-whisky-1.5237389.

Chapter 8

1. Antifolkhero, "Just Found Out Spotify Discontinued the Custom Running Playlist Feature," Reddit, March 20, 2018, https://www.reddit.com/r/spotify/comments/85uhpr/just_found_out_spotify_discontinued_the_custom/.

2. Antifolkhero, "Just Found Out."

3. Antifolkhero, "Just Found Out."

4. Owen Smith, Sten Garmark, and Gustav Söderström, Physiological Control Based on Media Content Selection, US Patent 10,209,950, filed December 22, 2016, and issued February 19, 2019.

5. See Maria Eriksson, Rasmus Fleischer, Anna Johansson, Pelle Snickars, and Patrick Vonderau, *Spotify Teardown: Inside the Black Box of Streaming Music* (Cambridge, MA: MIT Press, 2019), 121.

6. For an accessible introduction to recommender systems, see Dietmar Jannach, Markus Zanker, Alexander Felfernig, and Gerhard Friedrich, *Recommender Systems: An Introduction* (New York: Cambridge University Press, 2011); and Michael Schrage, *Recommendation Engines* (Cambridge, MA: MIT Press, 2020).

7. Paul Resnick and Hal R. Varian, "Recommender Systems," *Communications of the ACM* 40, no. 3 (1997): 56–58.

8. For a broader discussion, see Marc Andrejevic, *Automated Media* (London: Routledge, 2019), 8–10.

9. See the still existing website of the Simputer at http://www.simputer.org/.

10. N. Dayasindhu, "What Lessons Does the Simputer Hold for Made in India," Founding Fuel, January 13, 2017, https://www.foundingfuel.com/article/what-lessons-does-the-simputer-hold-for-make-in-india/.

11. Bruce Sterling, "The Year in Ideas: A to Z.; Simputer," *New York Times*, December 9, 2001, https://www.nytimes.com/2001/12/09/magazine/the-year-in-ideas-a-to-z-simputer.html.

12. To understand why the Simputer failed, see "Why did the Simputer Flop?," *Economic Times*, July 23, 2015, https://economictimes.indiatimes.com/news/science/why-did-the-simputer-flop/articleshow/48180974.cms.

13. See Margaret 'Espinasse, *Robert Hooke* (Berkeley: University of California Press, 1982), 117.

14. Catrine Tudor-Locke, Yoshiro Hatano, Robert P. Pangrazi, and Minsoo Kang, "Revisiting 'How Many Steps Are Enough?,'" *Medicine & Science in Sports & Exercise* 40, no. 7 (2008): 5537–5543.

Notes

15. The American social anthropologist Natasha Dow Schüll, in her studies of self-tracking cultures, labels these kinds of invisible monitoring *digital nudge* technologies. See Natasha Dow Schüll, "Data for Life: Wearable Technology and the Design of Self-Care," *BioSocieties* 11, no. 3 (2016): 317–333.

16. Gary Wolf, "The Data-Driven Life," *New York Times Magazine*, April 28, 2010, https://www.nytimes.com/2010/05/02/magazine/02self-measurement-t.html.

17. Wolf, "Data-Driven Life."

18. See Michel Foucault, *The Care of the Self*, trans. Robert Hurley (New York: Vintage Books, 1988); Foucault, "Self Writing," in *Ethics: Subjectivity and Truth*, trans. Robert Hurley, ed. Paul Rabinow (New York: The New Press, 1997), 208; and Foucault, "Technologies of the Self," in *Technologies of the Self: A Seminar with Michel Foucault*, ed. Luther H. Martin, Huck Gutman, and Patrick H. Hutton (Amherst: University of Massachusetts Press, 1988).

19. Kenneth R. Fyfe, James K. Rooney, and Kipling W. Fyfe, Motion Analysis System, US Patent 6,513,381, filed July 26, 2001, and issued February 4, 2003.

20. Wolf, "Data-Driven Life."

21. Dagognet, *Étienne-Jules Marey*, 15.

Chapter 9

1. Examples include the following: Jacob Kröger and Philip Raschke, "Is My Phone Listening In? On the Feasibility and Detectability of Mobile Eavesdropping," *IFIP Annual Conference on Data and Applications Security and Privacy* (Cham, Switzerland: Springer, 2019); Hamza Shaban, "Dust Isn't the Only Thing Your Roomba Is Sucking Up. It's Also Gathering Maps of Your House," *Washington Post*, July 15, 2017, https://www.washingtonpost.com/news/the-switch/wp/2017/07/25/the-company-behind-the-roomba-wants-to-sell-maps-of-your-home/; Niraj Chokshi, "Is Alexa Listening? Amazon Echo Sent Out Recording of Couple's Conversation," *New York Times*, May 25, 2018, https://www.nytimes.com/2018/05/25/business/amazon-alexa-conversation-shared-echo.html; Alyssa Newcomb, "Nest's Hidden Microphone Prompts Privacy Group to Call for FTC Action Against Google," *Fortune*, February 21, 2019, https://www.yahoo.com/lifestyle/nest-apos-hidden-microphone-prompts-221553176.html; Stacy Cowley, "Hold the Phone! My Unsettling Discoveries about How Our Gestures Online Are Tracked," *New York Times*, August 15, 2018, https://www.nytimes.com/2018/08/15/business/behavioral-biometrics-data-tracking.html; and Laura Hautala, "COVID-19 Contact Tracing Apps Create Privacy Pitfalls around the World," CNET, August 8, 2020, https://www.cnet.com/news/covid-contact-tracing-apps-bring-privacy-pitfalls-around-the-world/.

2. John Naughton, "'The Goal Is to Automate Us': Welcome to the Age of Surveillance Capitalism," *Guardian*, January 20, 2019, https://www.theguardian.com/technology/2019/jan/20/shoshana-zuboff-age-of-surveillance-capitalism-google-facebook.

3. Peter Cariani, "Some Epistemological Implications of Devices which Construct Their Own Sensors and Effectors," in *Towards a Practice of Autonomous Systems: Proceedings of the First European Conference on Artificial Life*, ed. Francisco J. Varela and Paul Bourgine (Cambridge, MA: MIT Press, 1992), 484–493.

4. Zhicheng Long, Bryan Quaife, Hanna Salman, and Zoltán N. Oltvai, "Cell-Cell Communication Enhances Bacterial Chemotaxis toward External Attractants," *Nature Scientific Reports* 7, no. 1 (2017): 12855.

5. See Peter Godfrey-Smith, *Other Minds: The Octopus, the Sea, and the Deep Origins of Consciousness* (New York: Farrar, Straus and Giroux, 2016).

6. See Peter Cariani, "Time Is of the Essence" (paper presented at the A Body of Knowledge—Embodied Cognition and the Arts conference, CTSA UC Irvine, Irvine, California, December 8–10, 2016).

7. See Cariani, "Some Epistemological Implications"; and W. Ross Ashby, *An Introduction to Cybernetics* (London: Chapman & Hall, 1956).

8. Cariani, "Epistemological Implications," 484.

9. Bateson's original context for this definition is that of an understanding of information. We can enlarge this here, however, by considering Cariani's claim that that sensing is about making distinctions or differences between potential states of the world. See Gregory Bateson, *Steps to an Ecology of Mind* (New York: Ballatine Books, 1972).

10. Peter Cariani in discussion with the author, June 2019. As neuroscientist, biologist, and cybernetician Cariani points out, the model of sensing—coordinating sensory states into neural firings and effecting (motor action)—is standard psychology. What makes the mix more intriguing and gives it a cybernetic basis is the fact that it "conceives of minds and brains in terms of purposive, purposeful systems." See Cariani, "Time Is of the Essence."

11. Malcolm Tatum, "What Is Machine Perception?," EasyTechJunkie, last modified February 20, 2021, https://www.easytechjunkie.com/what-is-machine-perception.htm.

12. See Richard O. Duda, Peter E. Hart, and David G. Stork, *Pattern Classification*, 2nd edition (London: John Wiley & Sons, 2001), 9–14.]

13. For early histories of AI, see J. McCarthy, M. L. Minsky, N. Rochester, and C. E. Shannon, "A Proposal for the Dartmouth Summer Research Project in Artificial

Intelligence, August 31, 1955," *AI Magazine* 27, no. 4 (2006): 12; Melanie Mitchell, *Artificial Intelligence: A Guide For Thinking Humans* (New York: Farrar, Strauss and Giroux, 2019), 29–59.

14. Douglas R. Hofstaedter, "On Seeing A's and Seeing As," *Stanford Humanities Review* 4, no. 2 (1995): 109–121.

15. Duda, Hart, and Stork, *Pattern Classification*, 5.

16. See David H. Hubel and Torsten N. Wiesel, "Receptive Fields, Binocular Interaction and Functional Architecture in the Cat's Visual Cortex," *Journal of Physiology* 160, no. 1 (1962): 106–154.

17. A *convolutional neural network* (CNN) is a neural network that has one or more convolutional layers. CNNs are used mainly for image processing, classification, segmentation, and other autocorrelated data. See Yann LeCun and Yoshua Bengio, "Convolutional Networks for Images, Speech, and Time Series," in *The Handbook of Brain Theory and Neural Networks*, ed. Michael Arbib (Cambridge, MA: MIT Press, 1995), 276–279.

18. The question of bias in machine learning systems is complex and multilayered. See Sidney Perkowitz, "The Bias in the Machine: Facial Recognition Technology and Racial Disparities," *MIT Case Studies in Social and Ethical Responsibilities of Computing*, February 6, 2021, https://mit-serc.pubpub.org/pub/bias-in-machine/release/1; and Bajwa, "What We Talk About When We Talk About Bias (A guide for everyone)." For a more general discussion, see Kate Crawford, *Atlas of AI* (New Haven: Yale University Press, 2021).

19. For a comparison between CNNs and the visual system, see Grace W. Lindsay, "Convolutional Neural Networks as a Model of the Visual System: Past, Present, and Future," *Journal of Cognitive Neuroscience*, February 6, 2020, 1–19, http://dx.doi.org/10.1162/jocn_a_01544.

20. For more on such nonconscious cognition, see N. Katherine Hayles, *Unthought: The Power of the Cognitive Non-Conscious* (Chicago: University of Chicago Press, 2017).

21. Nobert Wiener, *The Human Use of Human Beings: Cybernetics and Society* (London: Free Association Books, 1989), 47.

22. Richard F. Lyon, *Human and Machine Hearing* (Cambridge: Cambridge University Press, 2017), 6.

23. Jerry Lu, "Can You Hear Me Now? Far-Field Voice," *Towards Data Science* (blog), August 1, 2017, https://towardsdatascience.com/can-you-hear-me-now-far-field-voice-475298ae1fd3. For more technical detail, see Amit Chhetri, Philip Hilmes, Trausti Kristjansson, Wai Chu, Mohamed Mansour, Xiaoxue Li, and Xianxian Zhang, "Multichannel Audio Front-End for Far-Field Automatic Speech Recognition," in *2018*

26th European Signal Processing Conference (EUSIPCO) (New York: IEEE, 2018), 1527–1531.

24. See https://developer.amazon.com/alexa.

25. Kiran K. Edara, "Keyword Determinations from Conversational Data," US Patent 10,692,506, filed August 2, 2019, and issued June 23, 2020. As of 2019, Amazon applied for a new patent for *pre-wakeword speech processing*, which "describes a system in which speech is recorded, possibly in 10 to 30 second intervals, and processed for wakeword detection, possibly before being purged, and the next interval recorded."

26. Hidden Markov models are mathematical models based on the Markov model statistical concept. Named after the Russian mathematician Andrey Markov, in a Markov model, future values are statistically determined by present events and dependent only on the event immediately preceding. An HMM is used to capture "hidden" (nonobservable) information from an observable sequence of symbols—for example, predicting the weather (the so-called hidden variable) based on the color of clothing people might wear (the observed variable). HMMs follow a sequence of state changes (e.g., the weather today is sunny or cloudy) and build a model that includes the probabilities of each state progressing to another state. The goal is thus to determine the hidden parameters from the observable data. These models have long been used in speech recognition due to the dynamic, time-varying quality of speech and the ability to represent speech as a sequence of observations. For a technical explanation, see Lawrence Rabiner, "A Tutorial on Hidden Markov Models and Selected Applications in Speech Recognition," *Proceedings of the IEEE* 77, no. 2 (1989): 257–286.

27. Matt Day, Giles Turner, and Natalia Drozdiak, "Amazon Workers Are Listening to What You Tell Alexa," Bloomberg, April 10, 2019, https://www.bloomberg.com/news/articles/2019-04-10/is-anyone-listening-to-you-on-alexa-a-global-team-reviews-audio.

28. "The Boeing 737 Max MCAS Explained," *Aviation Week*, March 20, 2019, https://aviationweek.com/aerospace/boeing-737-max-mcas-explained.

29. Darryl Campbell, "Redline: The Many Errors that Brought Down the Boeing 737 Max," The Verge, May 2, 2019, https://www.theverge.com/2019/5/2/18518176/boeing-737-max-crash-problems-human-error-mcas-faa.

Chapter 10

1. Constant Nieuwenhuys, "The Great Game to Come," in *Theory of the Dérive and Other Situationist Writings on the City*, ed. Libero Andreotti and Xavier Costa (Barcelona: ACTAR, 1996), 62–63.

Notes

2. Johann Huizinga, *Homo Ludens: A Study of the Play-Element in Culture* (Boston: Beacon Press, 2009).

3. Constant Nieuwenhuys, "New Babylon," in *Gemeentenmuseum Den Haag*, exhibition catalog, 1974; available at https://www.readingdesign.org/new-babylon, Reading Design, accessed December 14, 2020.

4. Nieuwenhuys, "New Babylon."

5. Nieuwenhuys, "New Babylon."

6. See Edward C. Whitman, "SOSUS: The 'Secret Weapon' of Undersea Surveillance," *Undersea Warfare* 7, no. 2 (2005).

7. US Department of Defense, "Bugging the Battlefield," film, 1969, https://archive.org/details/gov.archives.arc.4524913, accessed August 10, 2019.

8. See Ann Finkbeiner, *The Jasons: The Secret History of Science's Postwar Elite* (London: Penguin, 2006).

9. John T. Correll, "Igloo White," *Air Force Magazine*, November 1, 2004, https://www.airforcemag.com/article/1104igloo/; Jan Philip Müller, "Radiophonics of the Vietnam War: A Collection," *Continent* 5, no. 3 (2016), https://web.archive.org/web/20200718125943/http://www.continentcontinent.cc/index.php/continent/article/view/257.

10. Correll, "Igloo White." See also Project CHECO Report, "Igloo White: July 1968–December 1969," HQ PACAF, Directorate, Tactical Evaluation, January 10, 1970.

11. Charles Q. Brown, "Fire Support Coordination Measures," in *Developing Doctrine for the Future Joint Force: Creating Synergy and Minimizing Seams* (Alabama: Air University Press, 2005), 53–63, http://www.jstor.org/stable/resrep13822.15.

12. David S. Alberts, John J. Garstka, and Frederick P. Stein, *Network Centric Warfare: Developing and Leveraging Information Superiority*, 2nd edition (Washington, DC: Command and Control Research Program/US Department of Defense, 2000), 2.

13. DARPA described the system in a matter-of-fact way in its concluding report for the DSN program: "A system of networked sensors can detect and track threats (e.g., winged and wheeled vehicles, personnel, chemical and biological agents) and be used for weapon targeting and area denial. Each sensor node will have embedded processing capability, and will potentially have multiple onboard sensors, operating in the acoustic, seismic, infrared (IR), and magnetic modes, as well as imagers and microradar. Also onboard will be storage, wireless links to neighboring nodes, and location and positioning knowledge through the global positioning system (GPS) or local positioning algorithms." See Chee-Yee Chong and Srikanta P. Kumar, "Sensor

Networks: Evolution, Opportunities, and Challenges," *Proceedings of the IEEE* 91, no. 8 (2003): 1247–1256; Arthur K. Cebrowski, VAdm, USN, and John H. Garstka, "Network Centric Warfare: Its Origin and Future," *Proceedings of the Naval Institute* 124, no. 1 (1998): 28–35.

14. Rich Haridy, "Plant Spies: DARPA's Plan to Create Organic Surveillance Sensors," New Atlas, November 21, 2017, https://newatlas.com/darpa-advanced-plant-technology-sensor-research/52292/. See also "Nature's Silent Sentinels Could Help Detect Security Threats," DARPA, November 17, 2017, https://www.darpa.mil/news-events/2017-11-17.

15. A classic textbook definition of such systems is as "a network of individual nodes that are able to interact with their environment by sensing or controlling physical parameters." See Holger Karl and Andreas Willig, *Protocols and Architectures for Wireless Sensor Networks* (Hoboken, NJ: John Wiley & Sons, 2007), 2.

16. Kazem Sohraby, Daniel Minoli, and Taieb Znati, *Wireless Sensor Networks: Technology, Protocols, and Applications* (Hoboken, NJ: John Wiley & Sons, 2007), 38–74.

17. Nieuwenhuys, "New Babylon."

18. See Lars Erik Holmquist et al., "Building Intelligent Environments with Smart-Its," *IEEE Computer Graphics and Applications* 24, no. 1 (2004): 56–64; Brett Warneke, Matt Last, Brian Liebowitz, and Kristofer S. J. Pister, "Smart Dust: Communicating with a Cubic-Millimeter Computer," *Computer* 34, no. 1 (2001): 44–51; and Keoma Brun-Laguna, Ana Laura Diedrichs, Diego Dujovne, Carlos Taffernaberry, Rémy Léone, Xavier Vilajosana, and Thomas Watteyne, "Using SmartMesh IP in Smart Agriculture and Smart Building Applications," *Computer Communications* 121 (2018): 83–90.

19. In the early 2000s, the US-based National Research Council issued a now well-cited report that argued: "EmNets will be implemented as a kind of digital nervous system to enable instrumentation of all sorts of spaces, ranging from in situ environmental monitoring to surveillance of battlespace conditions." See National Research Council, *Embedded, Everywhere: A Research Agenda for Networked Systems of Embedded Computers* (Washington, DC: National Academies Press, 2001).

20. See, for example, Diane J. Cook, Juan C. Augusto, and Vikramaditya R. Jakkula, "Ambient Intelligence: Technologies, Applications, and Opportunities," *Pervasive and Mobile Computing* 5, no. 4 (2009): 277–298; Uwe Hansmann, Lothar Merk, Martin S. Nicklous, and Thomas Stober, *Pervasive Computing Handbook* (Berlin: Springer Science & Business Media, 2000); M. Addlesee, R. Curwen, S. Hodges, J. Newman, P. Steggles, A. Ward, and A. Hopper, "Implementing a Sentient Computing System," *Computer* 34, no. 8 (2001): 50–56; and Adam Greenfield, *Everyware: The Dawning Age of Ubiquitous Computing* (Berkeley: New Riders, 2006).

21. For a discussion of remote sensing from a cultural studies perspective, see Jennifer Gabrys, *Program Earth: Environmental Sensing Technology and the Making of a Computational Planet* (Minneapolis: University of Minnesota Press, 2016).

22. For a comparison of fog, edge, and mist computing, see Yogesh Malik, "Internet of Things Bringing Fog, Edge and Mist Computing," *Yogesh Malik* (blog), September 20, 2017, https://yogeshmalik.medium.com/fog-computing-edge-computing-mist-computing-cloud-computing-fluid-computing ed965617d8f3.

23. Ying Chen, "COVID-19 Pandemic Imperils Weather Forecast," *Geophysical Research Letters* 47, no. 15 (2020), https://doi.org/10.1029/2020GL088613.

24. "ESRL/GSD Aircraft Data (AMDAR) Information," NOAA/ESRL/GSL Aircraft Data Web, last modified September 30, 2014, https://amdar.noaa.gov/FAQ.html.

25. "Reduction in Commercial Flights Due to COVID-19 Leading to Less Accurate Weather Forecasts," ScienceDaily, July 17, 2020, https://www.sciencedaily.com/releases/2020/07/200717101026.htm.

26. Thomas Lecocq, Stephen P. Hicks, Koen Van Noten, Kasper van Wijk, Paula Koelemeijer, Raphael S. M. De Plaen, Frédérick Massin et al., "Global Quieting of High-Frequency Seismic Noise Due to COVID-19 Pandemic Lockdown Measures," *Science* 369, no. 6509 (September 11, 2020): 1338–1343, https://science.sciencemag.org/content/369/6509/1338.

27. Lecocq et al., "Global Quieting," 1343.

28. See Dominique Rouillard, *Superarchitecture: Le futur de l'architecture, 1950–1970* (Paris: Editions de la Villette, 2004).

29. See Wolf D. Prix, *Get Off of My Cloud: Wolf D. Prix, Coop Himmelb(l)au: Texts 1968–2005* (Ostfildern-Ruit: Hatje Cantz, 2005).

30. Victoria Bugge Øye, "On Astroballoons and Personal Bubbles," e-flux, April 17, 2018, https://www.e-flux.com/architecture/positions/194841/on-astroballoons-and-personal-bubbles/.

31. See Chris Salter, *Entangled: Technology and the Transformation of Performance* (Cambridge, MA: MIT Press, 2010), 94–95.

32. See Malcolm McCulloch, *Ambient Commons: Attention in the Age of Ambient Information* (Cambridge, MA: MIT Press, 2013).

33. "What Is a Smart City?," Cisco, accessed April 12, 2020, https://www.cisco.com/c/en/us/solutions/industries/smart-connected-communities/what-is-a-smart-city.html.

34. See Halpern, *Beautiful Data*, 1–8.

35. Carlo Ratti and Matthew Claudel, *The City of Tomorrow: Sensors, Networks, Hackers, and the Future of Urban Life* (New Haven, CT: Yale University Press, 2016), 67.

36. See Peter H. Jones and Ove Nyquist Arup, *Ove Arup: Masterbuilder of the Twentieth Century* (New Haven, CT: Yale University Press, 2006).

37. Arup, *Digital Twin: Towards a Meaningful Framework* (London: Arup, November 2019), https://www.arup.com/perspectives/publications/research/section/digital-twin-towards-a-meaningful-framework.

38. See "Global Analyzer: Enables the Easy Identification of Trends and Mitigates Risk," Arup, accessed July 10, 2020, https://www.arup.com/projects/global-analyzer.

39. See "Neuron: AI Smart Building Console, Empowering Smart Buildings with a 'Digital Brain,'" Arup, accessed July 11, 2020, https://www.arup.com/projects/neuron.

40. William J. Holstein, "Virtual Singapore: Creating an Intelligent 3D Model to Improve Experiences of Residents, Business and Government," *Compass: The 3DExperience Magazine*, October 27, 2015, https://compassmag.3ds.com/8/Cover-Story/VIRTUAL-SINGAPORE.

41. See https://www.sidewalktoronto.ca/plans/quayside, accessed July 20, 2020.

42. Fussell, "City of the Future."

43. Blayne Haggart and Natasha Tusikov, "Sidewalk Labs' Smart-City Plans for Toronto Are Dead. What's Next?," The Conversation, May 8, 2020, https://theconversation.com/sidewalk-labs-smart-city-plans-for-toronto-are-dead-whats-next-138175.

44. Guangzhong Yang, *Body Sensor Networks* (New York: Springer, 2014), 102.

45. Yang, *Body Sensor Networks*, 3–4.

46. Dawn Nafus, "Introduction," in *Quantified: Biosensing Technologies in Everyday Life*, ed. Dawn Nafus (Cambridge, MA: MIT Press, 2016), xiii–xiv.

47. Nafus, "Introduction," xiv.

48. Nafus, "Introduction," xv.

49. Robert F. Service, "Can Sensors Make a Home in the Body?," *Science* 297, no. 5583 (August 9, 2002): 962–963, https://science.sciencemag.org/content/297/5583/962.

50. See Barbara T. Korel and Simon G. M. Koo, "A Survey on Context-Aware Sensing for Body Sensor Networks," *Wireless Sensor Network* 2, no. 8 (2010): 571–583.

51. Tim Harford, "Big Data: A Big Mistake?," *Significance* 11, no. 5 (2014): 14–19.

52. See "Predictive Modeling," OmniSci, accessed December 16, 2020, https://www.omnisci.com/technical-glossary/predictive-modeling.

53. Korel and Koo, "Survey on Context-Aware Sensing," 573.

54. Nieuwenhuys, "Great Game," 63.

Chapter 11

1. Joe Kamiya, "Conscious Control of Brain Waves," *Psychology Today* 1 (1968): 56–60. See also Joe Kamiya, "The First Communications about Operant Conditioning of the EEG," *Journal of Neurotherapy* 15, no. 1 (2011): 65.

2. Kamiya was not the only researcher interested in the conscious control of the brain. At about the same time, neuroscientist Barry Sterman at UCLA, inspired by reading about Kamiya's experiments, was investigating similar phenomena, this time using cats as opposed to human subjects. See Jim Robbins, *A Symphony in the Brain: The Evolution of the New Brain Wave Biofeedback* (New York: Grove Press, 2000). See also Erik Peper and Fred Shaffer, "Biofeedback History: An Alternative View," *Biofeedback* 46, no. 4 (2018): 80–85.

3. William Grey Walter was an American born, UK-based neurophysiologist who, in the 1950s and 1960s, performed pioneering work using EEG, as well as work in autonomous robotics. His 1953 book *The Living Brain* (London: Duckworth) also described his experiments in *flicker fusion*—using brainwaves to control stroboscopic light in order to explore altered states of consciousness.

4. Robbins, *Symphony in the Brain*, 38.

5. For more on Esalen, see Marion Goldman, *The American Soul Rush: Esalen and the Rise of Spiritual Privilege* (New York: NYU Press, 2012).

6. For the history of the movement, see Claudia X. Valdes and Philip Thurtle, "Biofeedback and the Arts: Listening as Experimental Practice" (paper presented at the REFRESH! conference, First International Conference on the Media Arts, Sciences and Technologies, Banff Center, Banff, Canada, September 29–October 4, 2005).

7. Richard Teitelbaum, "In Tune: Some Early Experiments in Biofeedback Music (1966–1974)," *Biofeedback and the Arts, Results of Early Experiments* (Vancouver: Aesthetic Research Center of Canada Publications, 1976), 36.

8. See, for example, Elizabeth Hinkle-Turner, "Women and Music Technology: Pioneers, Precedents and Issues in the United States," *Organised Sound* 8, no. 1 (2003): 31–34; Mirajana Prpa and Philippe Pasquier, "Brain-Computer Interfaces in Contemporary Art: A State of the Art and Taxonomy," in *Brain Art: Brain-Computer Interfaces for Artistic Expression*, ed. Anton Nijholt (Cham, Switzerland: Springer, 2019); and Anton Nijholt, "Introduction," in *Brain Art*, 1–29.

9. See Antoine Lutz, Lawrence L. Greischar, Nancy B. Rawlings, Matthieu Ricard, and Richard J. Davidson, "Long-Term Meditators Self-Induce High-Amplitude Gamma

Synchrony during Mental Practice," *Proceedings of the National Academy of Sciences* 101, no. 46 (2004): 16369–16373; and Christoph Koch, "Neuroscientists and the Dalai Lama Swap Insights on Meditation," *Scientific American*, July 1, 2013, https://www.scientificamerican.com/article/neuroscientists-dalai-lama-swap-insights-meditation/.

10. There is ongoing debate about the comparison of accuracy for wet versus dry electrodes.

11. This is the international standard for electrode placement. See R. W. Homan, "The 10–20 Electrode System and Cerebral Location," *American Journal of EEG Technology* 28, no. 4 (1988): 269–279.

12. See https://squaremile.com/gear/the-muse-headband/ and https://choosemuse.com/.

13. See http://neurosky.com/; http://o.macrotellect.com/; and https://brainbit.com/.

14. The original German is *Hirnspiegel*—literally, *brain mirror*.

15. See D. Millett, "Hans Berger: From Psychic Energy to the EEG," *Perspectives in Biology and Medicine* 44, no. 4 (2001): 522–542.

16. Vincent Walsh, "Brain Mapping: Faradization of the Mind," *Current Biology* 8, no. 1 (1998): 8–11, https://www.sciencedirect.com/science/article/pii/S0960982298700983.

17. Walsh, "Brain Mapping," 10.

18. There are technical issues with measuring the electrical change in the brain through TMS using EEG in terms of magnetic interference with the electrical field.

19. See Maria Konnikova, "Hacking the Brain: How We Might Make Ourselves Smarter in the Future," *Atlantic*, June 2015, https://www.theatlantic.com/magazine/archive/2015/06/brain-hacking/392084/.

20. See https://openbci.com.

21. See https://openbci.com.

22. See Carles Grau, Romuald Ginhoux, Alejandro Riera, Thanh Lam Nguyen, Hubert Chauvat, Michel Berg, Julià L. Amengual, Alvaro Pascual-Leone, and Giulio Ruffini, "Conscious Brain-to-Brain Communication in Humans Using Non-invasive Technologies," *PLoS ONE* 9, no. 8 (2014): e105225.

23. See Sydney Johnson, "Brainwave Headsets Are Making their Way into Classrooms—for Meditation and Discipline," *EdSurge*, November 14, 2017, https://www.edsurge.com/news/2017-11-14-brainwave-headsets-are-making-their-way-into-classrooms-for-meditation-and-discipline.

24. See Moheb Costandi, *Neuroplasticity* (Cambridge, MA: MIT Press, 2016), 41–44.

25. The most famous case of such natural neuroplasticity includes several accounts in the late neurologist Oliver Sacks's celebrated book *The Man Who Mistook His Wife for a Hat* (London: Picador, 2015).

26. Charles Lenay, Olivier Gapenne, Sylvain Hanneton, and Catherine K. Marque, "Sensory Substitution: Limits and Perspectives," in *Touching for Knowing: Cognitive Psychology of Haptic Manual Perception*, ed. Y. Hatwell, A. Streri, and E. Gentaz (Amsterdam/Philadelphia: John Benjamins Publishing Company, 2003): 276.

27. Benjamin W. White, Frank A. Saunders, Lawrence Scadden, Paul Bach-y-Rita, and Carter C. Collins, "Seeing with the Skin," *Perception & Psychophysics* 7, no. 1 (1970): 23–27.

28. Paul Bach-y-Rita, "Sensory Plasticity: Applications to a Vision Substitution System," *Acta Neurologica Scandinavica* 43, no. 4 (1967): 417–426.

29. Terry Gilliam, dir., *Brazil* (London: Embassy International Pictures, 1986).

30. There are numerous accounts of Bach-y-Rita's early TVSS experiments. See, for example, Paul Bach-y-Rita, "Tactile Vision Substitution: Past and Future," *International Journal of Neuroscience* 19, no. 1–4 (1983): 29–36; Paul Bach-y-Rita, Mitchell E. Tyler, and Kurt A. Kaczmarek, "Seeing with the Brain," *International Journal of Human-Computer Interaction* 15, no. 2 (2003): 285–295; and G. Guarniero, "Experience of Tactile Vision," *Perception* 3, no. 1 (1974): 101–104. For a more sociological take, see Mark Paterson, "Molyneux, Neuroplasticity, and Technologies of Sensory Substitution," in *The Senses and the History of Philosophy*, ed. Brian Glenney and Jose Filipe Silva (New York: Routledge, 2019); and Paterson, *Seeing with the Hands: Blindness, Vision and Touch after Descartes* (Edinburgh: Edinburgh University Press, 2016).

31. White et al., "Seeing with the Skin," 23.

32. See Bach-y-Rita, Carter C. Collins, Frank A. Saunders, Benjamin White, and Lawrence Scadden, "Vision Substitution by Tactile Image Projection," *Nature* 221, no. 5184 (1969): 963–964.

33. Yuri Danilov and Mitchell Tyler, "BrainPort: An Alternative Input to the Brain," *Journal of Integrative Neuroscience* 4, no. 4 (2005): 537–550.

34. Nicola Twilley, "Seeing with Your Tongue," *New Yorker*, May 8, 2017, https://www.newyorker.com/magazine/2017/05/15/seeing-with-your-tongue.

35. See, for example, neuroscientist David Eagleman's post about sensory substitution and his VEST technology at https://www.eagleman.com/research/sensory-substitution, accessed December 21, 2020.

36. National Research Council (US) Committee on Military and Intelligence Methodology for Emergent Neruophysiological and Cognitive/Neural Research in the

Next Two Decades, *Emerging Cognitive Neuroscience and Related Technologies* (Washington, DC: National Academies Press, 2008).

37. Jane Wakefield, "Brain Hack Devices Must Be Scrutinised, Say Top Scientists," BBC, September 10, 2019, https://www.bbc.com/news/technology-49606027.

38. Zachary Tomlinson, "Mind-Hunting: Could Your Brain be a Target for Hackers?," *Interesting Engineering*, November 20, 2018, https://interestingengineering.com/mind-hunting-could-your-brain-be-a-target-for-hackers; Daphne-Laprince Ringuet, "Brain-Hacking Is the Next Big Nightmare, So We'll Need Antivirus for the Mind," ZDNet, January 21, 2020, https://www.zdnet.com/article/brain-hacking-is-the-next-big-nightmare-so-well-need-anti-virus-for-the-mind/.

39. See Iris Coates McCall, Chloe Lau, Nicole Minielly, and Judy Illes, "Owning Ethical Innovation: Claims about Commercial Wearable Brain Technologies," *Neuron* 102, no. 4 (2019): 728–731.

40. Eliza Strickland, "Tech for Lucid Dreaming Takes Off—but Will Any of It Work?," *IEEE Spectrum*, July 14, 2017, https://spectrum.ieee.org/the-human-os/biomedical/devices/tech-for-lucid-dreaming-takes-off-but-will-any-of-it-work.

41. Penelope Green, "Sleep Is the New Status Symbol," *New York Times*, April 8, 2017, https://www.nytimes.com/2017/04/08/fashion/sleep-tips-and-tools.html.

42. Mathew Walker, *Why We Sleep: Unlocking the Power of Sleep and Dreams* (New York: Simon and Schuster, 2017), 19–20.

43. Marco Hafner, Martin Stepanek, Jirka Taylor, Wendy M. Troxel, and Christian van Stolk, *Why Sleep Matters—the Economic Costs of Insufficient Sleep: A Cross-Country Comparative Analysis* (Santa Monica, CA: RAND Corporation, 2016, https://www.rand.org/pubs/research_reports/RR1791.html.

44. *Sleeping Aids Market Research Report* (PSI Market Research, July 2020), https://www.psmarketresearch.com/market-analysis/sleeping-aids-market.

45. See https://sleepgadgets.io/eversleep/.

46. This "raiding" of talent by large companies such as Google, Apple, and Facebook has gone to an even stranger area: neuroscientists doing research on animal cognition in order to unlock the secrets of the animal mind for the training of self-driving cars and machine learning systems. See Sarah McBride and Ashlee Vance, "Apple, Google and Facebook are Raiding Animal Research Labs," Bloomberg, June 18, 2019, https://www.bloomberg.com/news/features/2019-06-18/apple-google-and-facebook-are-raiding-animal-research-labs.

47. G. Tononi and C. Cirelli, "Sleep and the Price of Plasticity: From Synaptic and Cellular Homeostasis to Memory Consolidation and Integration," *Neuron* 81, no. 1 (2014): 12–34. Sleep is also important in another aspect: it might be the last vestige

of respite from advanced capitalism. See Jonathan Crary, *24/7: Late Capitalism and the Ends of Sleep* (London: Verso, 2014).

48. See Vlad Savov, "The Sense Sleep Tracking Ball Is Nothing but a Lovely Alarm Clock," The Verge, July 26, 2016, https://www.theverge.com/circuitbreaker/2016/7/26/12283986/sense-sleep-tracker-review-alarm-clock.

49. See https://www.re-timer.com/the-science/light-therapy-for-sad-and-sleep/.

50. A 2011 poll from the National Sleep Foundation discovered that nine out of ten Americans utilized some kind of technological device one hour before they went to bed—devices that lead to some kinds of stressful feelings. See Michael Gradisar, Amy R. Wolfson, Allison G. Harvey, Lauren Hale, Russell Rosenberg, and Charles A. Czeisler, "The Sleep and Technology Use of Americans: Findings from the National Sleep Foundation's 2011 Sleep in America Poll," *Journal of Clinical Sleep Medicine* 9, no. 12 (December 15, 2013):1291–1299.

51. See Kelly Glazer Baron, Sabra Abbott, Nancy Jao, Natalie Manalo, and Rebecca Mullen, "Orthosomnia: Are Some Patients Taking the Quantified Self Too Far?," *Journal of Clinical Sleep Medicine* 13, no. 2 (February 15, 2017): 351–354. See also Emine Saner, "Why Sleeptrackers Could Lead to the Rise of Insomnia—and Orthosomnia," *Guardian*, June 17, 2019, https://www.theguardian.com/lifeandstyle/2019/jun/17/why-sleeptrackers-could-lead-to-the-rise-of-insomnia-and-orthosomnia.

52. See https://www.mindmachines.com/. See also Michael Hutchinson, *Mega Brain: New Tools and Techniques for Brain Growth and Mind Expansion* (Scotts Valley, AZ: CreateSpace Independent Publishing Platform, 2013).

53. Carl G. Jung, *The Practice of Psychotherapy*, trans. R. F. C. Hull (Florence: Taylor and Francis, 2014), 142.

54. See http://www.lucidity.com/.

55. See https://www.media.mit.edu/projects/theme-engineering-dreams/overview.

56. See https://www.media.mit.edu/projects/sleep-creativity/overview/.

57. See http://www.adamjhh.com/dormio.

58. *Hypnagogia* is an inexplicable liminal zone between being awake and asleep, involving an unstable mix of hallucinations, spatial-temporal distortions, and sensory illusions. It's essentially the mind producing phantasms. From Edison and Poe to Dali and Nikolai Tesla, artists, writers, inventors, and other creative individuals attempted to understand the experience of hypnagogia by taking *micronaps* (sleep periods lasting less than a quarter of a second) with steel balls, keys, and other objects in their hands. When the subjects fell asleep, their grip on the objects loosened and they fell, clattering to the floor or onto a special steel plate and thus immediately awakening them. See Adam Haar Horowitz et al., "Dormio: Interfacing with

Dreams," *Extended Abstracts of the 2018 CHI Conference on Human Factors in Computing Systems* (New York: ACM, 2018), 1–10.

59. For a scientific critique, see Catherine Offord, "Scientists Engineer Dreams to Understand the Sleeping Mind," *The Scientist*, December 1, 2020, https://www.the-scientist.com/features/scientists-engineer-dreams-to-understand-the-sleeping-brain-68170.

60. Douglas Trumbull, dir., *Brainstorm* (Moreno Valley, CA: JF Productions, 1983).

61. Katherine Bigelow, dir., *Strange Days* (Los Angeles: Lightstorm Entertainment, 1995).

62. Eben Harrell, "Neuromarketing: What You Need to Know," *Harvard Business Review*, January 23, 2019, https://hbr.org/2019/01/neuromarketing-what-you-need-to-know.

63. See Michael M. Maharbiz, Rikky Muller, Elad Alon, Jan M. Rabaey, and Jose M. Carmena, "Reliable Next-Generation Cortical Interfaces for Chronic Brain–Machine Interfaces and Neuroscience," *Proceedings of the IEEE* 105, no. 1 (2016): 73–82.

64. See https://neuralink.com/.

65. "Six Paths to the Nonsurgical Future of Brain-Machine Interfaces," DARPA, May 20, 2019, https://www.darpa.mil/news-events/2019-05-20.

66. See Jonna Brenninkmeijer, *Neurotechnologies of the Self: Mind, Brain and Subjectivity* (London: Palgrave Macmillan, 2016), 378.

67. Brenninkmeijer, *Neurotechnologies*.

68. See Dan Lloyd, "Outsourcing the Mind," *American Scientist* 97, no. 4 (July–August 2009), 340; Sam Kriss, "You Think with the World, Not Just Your Brain," *Atlantic*, October 13, 2017, https://www.theatlantic.com/science/archive/2017/10/extended-embodied-cognition/542808/; Jeffrey M. Schwartz and Rebecca Gladding, *You Are Not Your Brain: The 4-Step Solution for Changing Bad Habits, Ending Unhealthy Thinking, and Taking Control of Your Life* (New York: Penguin, 2011); and Hilary Putnam, "The Meaning of Meaning," in *Mind, Language and Reality; Philosophical Papers, Volume 2* (Cambridge: Cambridge University Press, 1975), 215–271.

69. Alva Noë, *Out of Our Heads: Why You Are Not Your Brain, and Other Lessons from the Biology of Consciousness* (New York: Hill and Wang, 2009), 18–19.

70. Andy Clark, "Re-inventing Ourselves: The Plasticity of Embodiment, Sensing, and Mind," *Journal of Medicine and Philosophy* 32, no. 3 (2007): 263–282.

71. Andy Clark, *Natural-Born Cyborgs: Minds, Technologies, and the Future of Human Intelligence* (Oxford: Oxford University Press, 2004), 6–7. See also Andy Clark and David Chalmers, "The Extended Mind," *Analysis* 58, no. 1 (1998): 7–19.

72. Clark, *Natural-Born Cyborgs*, 7.

73. Clark, *Natural-Born Cyborgs*, 138. The concept of an agent-world circuit comes from Clark as well. See Clark, "Re-inventing Ourselves," 266.

74. British sociologist of science Andrew Pickering has an interesting take on EEG. Such a technology acts for Pickering as a tool for self-experimentation; it enables us to achieve a performative model of ourselves. This implies that something or someone is not a fixed object but emerges in and over time. For Pickering, technologies such as wearable EEGs don't reveal some kind of objective, scientific self; they are techniques that enable us to work on and change us—part of an exploratory loop. There is no exterior fixed thing, for instance, like the brain, that can be isolated from the ever-changing environmental context we find ourselves in. It adapts and, along with this, the self changes as well. See Pickering, *Cybernetic Brain*, 384–385.

Epilogue

1. See Dugald J. M. Thomson and David R. Barclay, "Real-Time Observations of the Impact of COVID-19 on Underwater Noise," *Journal of the Acoustical Society of America* 147, no. 5 (2020): 3390–3396.

2. See C. Rutz, M.-C. Loretto, A. E. Bates, S. C. Davidson, C. M. Duarte, W. Jetz, and M. Johnson et al., "COVID-19 Lockdown Allows Researchers to Quantify the Effects of Human Activity on Wildlife," *Nature Ecology and Evolution* 4, no. 9 (2020): 1156–1159.

3. Rutz et al., "COVID-19 Lockdown."

4. N. S. Diffenbaugh, C. B. Field, E. A. Appel, I. L. Azevedo, D. D. Baldocchi, M. Burke, J. A. Burney et al., "The COVID-19 Lockdowns: A Window into the Earth System," *Nature Reviews Earth and Environment* 1, no. 9 (2020): 470–481.

5. Jet Propulsion Laboratory, California Institute of Technology, "ECOSTRESS," accessed September 20, 2020, https://ecostress.jpl.nasa.gov/instrument.

6. "NASA Funds Eight New Projects Exploring Connections between the Environment and COVID-19," NASA, September 3, 2020, https://www.nasa.gov/feature/esd/2020/new-projects-explore-connections-between-environment-and-covid-19.

7. Roberto Molar Candanosa, "These New Sensors Can Detect Coronavirus Particles on Your Breath, Instantly," News@Northeastern, July 17, 2020, https://news.northeastern.edu/2020/07/17/could-this-new-gas-sensor-help-researchers-test-regularly-for-the-coronavirus-in-the-air/; Wenyu Wang, Karim Ouaras, Alexandra L. Rutz, Xia Li, Magda Gerigk, Tobias E. Naegele, George G. Malliaras, and Yan Yan Shery Huang, "Inflight Fiber Printing toward Array and 3D Optoelectronic and Sensing Architectures," *Science Advances* 6, no. 40 (2020); Jordi Laguarta, Ferran Hueto, and Brian

Subirana, "COVID-19 Artificial Intelligence Diagnosis Using Only Cough Recordings," *IEEE Open Journal of Engineering in Medicine and Biology* 1 (September 2020): 275–281; F. Laghrib, S. Saqrane, Y. El Bouabi, A. Farahi, M. Bakasse, S. Lahrich, and M. A. El Mhammedi, "Current Progress on COVID-19 Related to Biosensing Technologies: New Opportunity for Detection and Monitoring of Viruses," *Microchemical Journal* 160 (January 2021): 105606; and Badriyah Alhalaili, Ileana Nicoleta Popescu, Olfa Kamoun, Feras Alzubi, Sami Alawadhia, and Ruxandra Vidu, "Nanobiosensors for the Detection of Novel Coronavirus 2019-nCoV and Other Pandemic/Epidemic Respiratory Viruses: A Review," *Sensors* 20, no. 22 (2020): 6591.

8. Kevin Jiang, "How COVID-19 Causes Loss of Smell," Harvard Medical School, July 24, 2020, https://hms.harvard.edu/news/how-covid-19-causes-loss-smell.

9. David H. Brann, Tatsuya Tsukahara, Caleb Weinreb, Marcela Lipovsek, Koen Van den Berge, Boying Gong, Rebecca Chance et al., "Non-neuronal Expression of SARS-CoV-2 Entry Genes in the Olfactory System Suggests Mechanisms Underlying COVID-19-Associated Anosmia," *Science Advances* 6, no. 31 (2020).

10. John Geddie and Aradhana Aravindan, "Singapore Plans Wearable Virus-Tracking Device for All," Reuters, June 4, 2020, https://www.reuters.com/article/us-health-coronavirus-singapore-tech-idUSKBN23C0FO; David Horsley and Richard J. Przybyla, "Ultrasonic Range Sensors Bring Precision to Social-Distance Monitoring and Contact Tracing," *Electronic Design*, September 2, 2020, https://www.electronicdesign.com/industrial-automation/article/21139839/ultrasonic-range-sensors-bring-precision-to-socialdistance-monitoring-and-contact-tracing; and Justin McCurry, "Japan Shop Deploys Robot to Check People Are Wearing Face Masks," *Guardian*, November 16, 2020, https://www.theguardian.com/world/2020/nov/16/japan-shop-deploys-robot-to-check-people-are-wearing-face-masks.

11. See S. Lalmuanawma, J. Hussain, and L. Chhakchhuak, "Applications of Machine Learning and Artificial Intelligence for Covid-19 (SARS-CoV-2) Pandemic: A Review," *Chaos, Solitons & Fractals* 139 (October 2020).

12. See Amanda E. Bates, Richard B. Primack, Paula Moraga, and Carlos M. Duarte, "COVID-19 Pandemic and Associated Lockdown as a 'Global Human Confinement Experiment' to Investigate Biodiversity Conservation," *Biological Conservation* 248 (August 2020): 108665.

13. See Zuboff, *Age of Surveillance Capitalism*.

14. See the reference from Eleanor Gibson (Gibson's wife and collaborator): E. J. Gibson and A. D. Pick, *An Ecological Approach to Perceptual Learning and Development* (Oxford: Oxford University Press, 2000); and James J. Gibson, *The Ecological Approach to Visual Perception: Classic Edition* (New York: Psychology Press, 2015). Interestingly, Gibson's theory of visual perception critiqued the long-standing idea since Newton that vision is about reproducing the visual world directly on the retina. Instead,

information about the environment is conveyed through ambient (surrounding) light that is accessible directly to the eye rather than being based upon visual cues (or clues) from the retina that have to be interpreted. This theory is also importantly directly tied to Gibson's critique of psychophysics. As he writes, "I failed to distinguish between stimulation proper and stimulus information, between what happens at passive receptors and what is available to active perceptual systems. Traditional psychophysics is a laboratory discipline in which physical stimuli are applied to an observer. He is prodded with controlled and systematically varied bits of energy so as to discover how his experience varies correspondingly. This procedure makes it difficult or impossible for the observer to extract *invariants* (information that does not change) over time. Stimulus prods do not ordinarily carry information about the environment." J. Gibson, *The Ecological Approach to Visual Perception: Classic Edition*, 141.

15. See "Transmission of SARS-CoV-2: Implications for Infection Prevention Precautions," WHO, July 9, 2020, https://www.who.int/news-room/commentaries/detail/transmission-of-sars-cov-2-implications-for-infection-prevention-precautions.

16. Science studies scholar Bruno Latour claims that reading a modern newspaper continually confronts us with what he calls "hybrids," imbroglios of science, politics, economy, law, religion, technology, fiction. "If reading the daily paper is modern man's form of prayer, then it is a very strange man indeed who is doing the praying today while reading about these mixed-up affairs. All of culture and all of nature get churned up again every day." See Bruno Latour, *We Have Never Been Modern*, trans. Catherine Porter (Cambridge, MA: Harvard University Press, 1993), 2.

17. Hayek, *Sensory Order*, 7.

18. Friedrich August Hayek, *Kinds of Order in Society* (Chicago: Institute for Humane Studies, 1975), 2.

19. Hayek, *Sensory Order*, 49–50.

20. Frank Rosenblatt, "The Perceptron: A Probabilistic Model for Information Storage and Organization in the Brain," *Psychological Review* 65, no. 6 (1958): 386–408.

21. Rosenblatt, "The Perceptron," 386.

Index

2001: A Space Odyssey (film), 9
5G Networks, 159

Absolute threshold, 22
Acceleration, 49, 53–55
  and centripetal force, 49
  definition in calculus, 49–51
  and gravity, 53
Accelerograph, 50
Accelerometer, 45. *See also* Controller; Sensors; Wiimote (or Wii Remote)
  3D, 171
  early history of, 51–53
  in everyday life, 50
  mechanics of, 50–51
  and MEMS technologies, 55
  and physics principles, 49, 52–54, 58
  Ryan's explanation of, 48–49
  use for air bags, 53
  use in musical instruments, 55–58
Accelerometry, 175
Accuracy (sensor measurement), 65
Actigraphy (sleep tracking), 235. *See also* Sleep
Actuator, 55, 128, 133, 182, 228, 230–231. *See also* Affector
Adidas, 176
Adria, Ferran and Albert, 150
Advanced Plant Technology (APT), 199
Affector, 180

Affordances, 113, 124, 277n37. *See also* Ecological theory of perception; Gibson, James J.
Afterimages, 16
AGE, 66–67. *See also* Mattel Power Glove
AI: More Than Human, 110, 123
Air pantographe, 32
Aircraft Meteorological Data Relay (AMDAR), 204
Alexa (Amazon technology), 6, 190
Algorithms, 6, 8, 36, 85, 108, 143, 176, 191, 214, 234–235. *See also* Bias, in AI algorithms and machine learning
  definition of, 58
  learning patterns, 184
  and Spotify, 168–169
  use in E-Nose software, 159
  use of in Kinect, 73, 76–77
Alien Zoo, 88
Allen, Paul, 57
Alpha waves, 101, 219–220, 222, 226, 243–244. *See also* Brainwaves
Alter (robot), 107–108, 120–124
Altman, Mitch, 67
Amazon, 8
  eavesdropping on conversations, 190–192
  Echo, 190, 192
  use of human listeners, 192
Ambient, 208

Ambient city, 195
Ambient intelligence, 202
American Telephone and Telegraph Corporation, 196–197. *See also* Bell Labs
Amida Simputer, 169–171
 failure of, 171
 first use of accelerometer in, 170
 and international publicity, 169
Analog Devices, 72–73
Anderson, Laurie, 57
Anechoic chamber, 100
Animal Magnetism (hypnosis), 16
Anosmia, 250
Anthropause, 248
Apollon musagète, 45
Apple, 8, 28, 36, 139, 145, 170, 176,
Apple Exercise Lab, 37
Apple Watch, 35, 38, 176,
AREA15 (Las Vegas), 88
Arrays (sensors), 5, 27, 46, 57, 147, 158, 160, 197, 201, 203, 230
Arte Programmata, 13, 95
Artificial intelligence, 7–8, 90, 109–110, 183, 213. *See also* Artificial neural network; Machine learning
 Kinect's use of, 73
 use of in computer vision, 187
Arup (architectural engineering firm), 6, 52, 209. *See also* Arup, Ove
Arup, Ove, 209
Art and technology program of LACMA, 97–100
Artforum, 98
Artificial life (A-Life), 109–113, 123
Artificial neural network, 114. *See also* Artificial intelligence; Hayek, Friedrich
 comparison to real neurons, 116–117
 and connectionism, 119
 definition of, 116–121, 188
 Hayek's description of, 117–119
 McCulloch and Pitts description of, 116

"Artwork in the Age of Technical Reproducibility, The," 92
Astroballoon, 206
Atari, 78
Atomizer, 1, 6
Automatic Speech Recognition (ASR), 191
Autonomous sensor network, 113–114
Autonomy, 109–11
Autopoiesis, 110

Bach, Richard, 219
Bach-y-Rita, Paul, 230–232
Ballmer, Steve, 74
Barbican Centre, 110, 123
Bateson, Gregory, 182
Bauhaus, 90
Bayesian estimation, 130, 213, 278n3
Beamforming, 190
Behavior, 19, 22, 108, 110, 119, 166, 206, 220, 225, 229, 242, 248. *See also* Extraction
 and autopoiesis, 110
 and cybernetics, 95, 102
 between humans and machines, 113, 123
 human, 94, 168, 175
 machines exhibiting, 122, 148, 189, 255
 in mathematical models, 214
 in organisms, 180
 self-tracking, 174
 in sensed spaces, 196
 surveillance capitalism's capture of, 8–9
Behavioral biometrics, 6
Bell Labs, 54, 99, 197
 Nine Evenings development, 96, 99
Béhar, Yves, 236
Benjamin, Walter, 92, 94
 experience of shock in urban life, 92
 sensory perception and technology, 92

Index

Bias, 1
  in AI algorithms and machine learning, 37, 116, 266n53, 287n18
  in data sets, 183–184
  in sensors, 1, 6,
  in face recognition, 139
  in sensory science experiments, 154–155
Bielecki, Bob, 57
Big data, 7–8, 39
Bigelow, Katherine, 241
Binaural Sound, 24, 238, 280n23
Biosensors, 157–158, 250. *See also* Sensors
  use in body sensor networks, 211–213
Bit rate, 184
Blade Runner, 9
Blaupunkt, 136–137
Blob detection, 85
BMW, 131, 135, 138, 144–145
Body as test subject, 214
Body area network (BAN), 211
Body sensor network (BSN), 211–213
Boeing, 192. *See also* Sensors
  angle of attack sensors failure, 192
  design errors for 737 Max plane, 192–193
  Lion Air and Ethiopian Air disasters, 192
Book of Knowledge of Ingenious Mechanical Devices, 9. *See also* Al-Jazari, Ismail
Boulez, Pierre, 56
Brain-computer interface (BCI), 6, 222, 225, 228, 232, 234, 240, 242
Brain hacking, 228–229, 233
BrainPort, 232
Brainstorm, 241
Brainwaves, 99, 101, 225, 238, 244. *See also* Alpha waves
  workers' being monitored in China, 6
Brenninkmeijer, Jonna, 243

Building information modeling (BIM), 5, 202, 208
Bundeswehr University, 127
Burnham, Jack, 95
Burning Man, 88

Cage, John, 99, 221
Calasso, Roberto, 44
Calibration, 65–66
Campbell Soup Company, 156
Cargill, 156
Caspersen, Dana, 46, 49–50
CCD (charge-coupled device), 27
Centre Pompidou, 103
Cephalopods, 180
Chalmers, David, 245
Chaotic processes (physics), 112, 122
Chemoresistor, 159
Chemotaxis, 180
Chen, Ying, 204
Childs, Lucinda, 99
Chronophotography, 30
Chronoscope, 18, 29, 33, 35
Cisco Systems, 208
Clark, Andy, 245
Classification (mathematical), 118, 159–160, 183, 214, 255. *See also* Artificial neural network; Bias
  patterns, 185–186
Closed-loop (engineering), 98
Co-constitutive, 245
Cold War, 94, 197
Communication (in cybernetics), 94, 102
Commutator, 230
Computer music, 56
  and real time, 47
Connectionism, 119. *See also* Artificial neural network
Consciousness, 17, 19–20, 22, 32, 44, 91–92, 100–101, 108, 145, 220, 225
  studies of, 222
Consumer Electronics Show (CES), 64

Consumer Reports, 77
Contact tracing, 1, 5, 79. 250–251. *See also* COVID-19 pandemic
Context awareness, 213
Control (in cybernetics), 94, 99, 102–105, 135, 139, 246, 278n2
Controller, 23, 46, 56, 58
  game, 50, 61–63, 65, 67, 69, 71–75, 78–79, 131, 171
  in VR, 23–25
Convolutional neural network (CNN), 187. *See also* Artificial neural network
Coop Himmelb(l)au, 206–207. *See also* Astroballoon
Cornell University, 35, 255
Cooper, Muriel, 32
Coupling (physics), 29, 111–112
COVID-19 pandemic, 85, 150
  biologging initiative, 248
  and contact tracing, 5
  creation of new sensors to combat, 249–250
  effect on smell, 250
  entanglement of biological and technical through, 252–253
  global quieting, 205–206
  and loss of weather observations due to, 204–205
  reduction of ocean noise due to, 247–248
  uncertainty due to, 210
"Cybernated Art" (Paik), 95
Cybernetics, 102–104, 189. *See also* Behavior; Communication; Control; Feedback
Cyborg, 5, 245
  origin of, 260n6

Dassault, 210
Data exhaust, 213
Data glove (VPL), 66–69. *See also* VPL (Silicon Valley Company)
Data science, 7

da Vinci, Leonardo, 171
Defense Advanced Research Projects Agency (DARPA), 130, 198–199, 242, 278n5
Degrees of freedom, 25
  six dof sensor, 25, 27, 236
Dehydrator, 7
Denso (Japanese robot corporation), 112
Depth map, 76–77
Design method, 26
  psychophysics as, 26–27
Dhvani (text to speech system), 169
Diacetyl, 153
Dickmanns, Ernst, 127
  development of dynamic vision concept, 128–129
  early research, 127
  and perception-action cycle, 130
  and self-driving cars, 128–130
  use of sensors, 127–130
Differential equation, 120–121
  definition of, 163n15
Digital Twins (software from Arup), 209
Disney, Walt, 86
Disney (corporation), 87–89
Distributed Sensor Networks project, 199
Documenta (art exhibition), 95
DreamLab (MIT research lab), 240–241
  and Dormio dream sensing, 240–241
  and relationship to *Inception* film, 241
DreamLight, 240
Dreams, 239. *See also* LaBerge, Stephen; Muscle atonia
  Jung's understanding of, 239
  and lucid dreams, 240–241
  stages of, 239
Dreamscape Immersive, 87–88
Dreem2, 236
Driver assist systems, 134
Drones, 1, 232
Duck Hunt (game), 63
Duchamp, Marcel, 90
Dummy head (psychoacoustics), 137

Index

Dynamics (in physics), 111–113, 119, 121–124
Dynastream, 176

ECOSTRESS, 249
E.coli bacteria, 180
Ecological theory of perception, 252
Edge computing, 202, 204
Edge detection, 73, 85, 157
Ego-vehicle, 135
Eidos:Telos (Ballet), 43–46, 50, 53, 56
Electrocardiogram (ECG), 143
Electroencephalogram (EEG), 219
  commercialization into consumer wearables, 224–225
  history of, 223, 226–227
  operation of, 223–224
Electromagnetic induction, 227
Electronic Arts (game company), 19
"Emerging Cognitive Neuroscience and Related Technologies" (report), 232
Eno, Brian, 87
E-nose
  operation of, 159
  research on arthropods' sense of smell, 159
  use of in food industry, 158
  and use of software for brain, 159
Environment soothing, 143
EPSON teamLab Borderless
  description of, 83–84
  historical reference points for, 89–90
  and use of sensors, 85–86
Equinix NY4 Data Center, 5
E-tongue, 147–149, 151, 157–158, 160–161
Eversley, Frederick, 98
Everyone, Everything, Everywhere, All the Time (E3A), 210
Experience, 18, 23–25, 49, 53, 148, 174, 214, 222, 254
  aesthetic, 11, 145
  and immersion, 29
  produced by Spotify Running, 166–169
  produced by brainwaves, 219–220, 223, 243
  as raw material, 8
  and the self, 243
  sensory, 12, 18–19, 21, 31, 254
  in video games, 62–63, 71–72
  Wundt's measurement of, 32–33
Experience economy, 7, 87–88
Experimental psychology, 13, 18
  Wundt's founding of, 32–33
Experimental self-observation, 34. *See also* Wundt, Wilhelm
Experiments in Art and Technology (E.A.T.), 100
Exposition internationale du Surréalisme, 90
Extended Mind Theory, 245–246
Extraction, 8, 203
  feature, 185–186, 203

FAANG (Facebook, Apple, Amazon, Netflix, Google), 8, 139, 179
Facebook Reality Labs, 19, 26
Facial recognition, 6, 210
  in cars, 139–140
  use of in Kinect, 73
Faraday, Michael, 227–228
Feature bloat, 193
Features, 128, 177. *See also* Extraction, feature
  definition of, 186–188
  detectors, 186
Fechner, Gustav, 11–23. *See also* Psychophysics
  critique of materialism, 19
  discovery of psychophysics, 22, 39–40
  illness, 15–16
  and just noticeable difference, 21
  revision of Weber's law, 21, 263n15
  theories applied to VR industry, 25–29

Feedback (in cybernetics), 94–97, 101–103, 105, 122, 132–133, 136, 144, 181, 193, 203, 211, 240. *See also* Control (in cybernetics); Cybernetics
Feynman, Richard, 54
   proposal for nanotechnology, 55
Fitbit, 36, 38–39, 50, 175–177, 236–237
Fitness, 173. *See also* Fitbit; Running (Spotify app)
   first digital diary for Nokia phone, 173
   tracking of with sensors, 2, 37, 50, 166, 168, 172 174–175, 177, 185, 223, 235
Flexion (finger bending), 67
Fog computing, 202, 204
Forsythe, William, 43–45
Foster, Norman, 52
Foucault, Michel, 174
Franke, Maxim, 45
Frankenstein, 9
Frankfurt Ballet, 43
Freud, Sigmund, 116
Fun Palace, 101–103
Fyfe, Ken (engineer), 175–177

Gait, 75, 176
Galvanic Vestibular Response (GVS), 133–134
Game Boy Color, 72
Ganzfeld effect, 100. *See also* Irwin, Robert; Turrell, James
Garmin, 176
Garrett Corporation, 98, 100
Gastronomic engineering, 7
Gastronomy, 160
  molecular, 160
  neuro-, 152
Geminoid (robot), 107, 112
Gesture recognition, 77–78
Gibson, James J., 252, 277n37
Gilmore, James, 87
Glia, 236

Global Analyzer (software from Arup), 209
Global human confinement experiment, 251
Goebbels, Joseph, 93
Google, 130, 175, 188
  chief economist Hal Varian's understanding of monitoring, 8
  as part of FAANG corporations, 8
  and Sidewalk Lab's Toronto project, 208, 210
  and surveillance capitalism, 10, 179,
  use of neural networks, 115, 121
Google Home assistant, 6
Graphic method, 31–32, 36
G-force, 53
"Great Game to Come, The," 195. *See also* Nieuwenhuys, Constant
Gruppo T, 95–96
Guitar Hero, 62
Gurner, Asaf, 64

Happenings (art event), 78, 87. *See also* Kaprow, Allan
Haraway, Donna, 108
Harter Raum (Hard Space), 206
Harvard Business Review, 87
Hatano, Yoshiro, 172
Hayek, Friedrich, 117–120. *See also Sensory Order, The*
  and classification, 118
  concept of order, 253
  description of sensory order, 118
  Ikegami's interest in, 120, 123
  and machines, 254–255
Heart-rate monitor, 6, 140, 260n7
Hebb, Donald, 120
Hebbian learning, 120. *See also* Hebb, Donald
Helmholtz, Hermann von, 31
Helmholtz resonator, 139
Hidden Markov model (HMM), 191, 213, 288n26

Index

Hollywood, 94
Hoftstadter, Douglas, 183
Homeostasis, 180
Homo Ludens, 195
Hong-Kong-Zhuhai-Macao Bridge, 5, 260n7
Hooke, Robert, 172
House of No Return, 88. *See also* Meow Wolf
Howes, David, 157
HTC Vive, 25
Hubel, David, 186
Huizinga, Johann von, 195–196
Human senses, 4, 7, 86, 106, 119
  and COVID-19, 250
  engineering and replacement of, 153, 155, 157
  entangled with machines, 9–10, 16, 40
  as extensions, 30
  Fechner's interest in, 26
  limitation of, 94
  as machine input, 28, 30, 37
  Marey's doubt of, 31, 35
HVAC (heating, ventilation and air conditioning), 6, 209
Hybrids, 253
Hydrophone, 197–198, 247

IBM, 133
Igloo White, 198
Ikegami, Takashi, 107–108
  and chaotic processes, 122
  criticism of robotics, 112
  discussion of sensors, 123
  and dynamics, 111
  maximalist approach, 112–113
  use of spiking neural network, 120–121
  work with Alife, 109
IMAX, 89
Inami, Masahiko, 133
Information (computing), 54, 59, 94, 102, 110–112, 114, 122, 130, 177, 182, 189, 214, 230, 255

Inner perception, 34
Input device, 48
Inside out sensors, 25
Integrated circuit, 95
Internet of Things, 209
iPhone, 170, 177
IRCAM, 56
Instrument (musical), 45, 55–58, 64, 102, 131
Instrument (scientific), 18–19, 30–40, 91, 156–157, 177–178
Intercontinental Ballistic Missile (ICBM), 52
International Space Station, 249
Interval Research Corporation, 57
Irwin, Robert, 100–101
Ishiguro, Hiroshi, 107–108, 111–112

Jaguar (car), 139
Jasons, 198
Jazari, Ismail al-, 9
Jefferson, Thomas, 172
Jet propulsion, 52, 59
Jet Propulsion Laboratory (JPL), 52, 197
Jobs, Steve, 169
Just noticeable difference (JND), 11, 21, 23
Just Noticeable Difference (JND) (installation), 13

Kalman filter, 130
Kamiya, Joe, 219–220, 223, 226, 239
Kaprow, Allan, 79
Karakuri Ningyo, 9
Kelly, Kevin, 174
Kickstarter, 238, 240
Kinect, 72–79
  and machine learning, 75
  and natural interfaces, 74
  operation through motion tracking, 76
  use of computer vision in, 75–76
Kirby Tilt 'n' Tumble (video game), 72

Klüver, Billy, 96, 99
Kracauer, Siegfried, 93
Kraft (food conglomerate), 158
Kusama, Yayoi, 89
Kymographion, 18, 31, 36, 38

LaBerge, Stephen, 240
Lahne, Jacob, 157
Langton, Christopher, 109
Lanier, Jaron, 67
Laser disc, 56, 268n17
Latency, 26
Lawrence UC Berkeley Labs, 46–47
LCD screens, 61
LEDs, 6, 36, 85, 110, 140
Light harp, 64
LinkedIn, 16–17, 19
Lion Air, 192
Lipids, 147
Liquid nitrogen, 7, 151
Littlewood, Joan, 101–103. *See also* Fun Palace
Live signal processing, 45
Livingston, Jane, 97
Logarithms, 263n16
  Fechner's use of, 22
Logic gate, 116–117
"Logical Calculus of the Ideas Immanent in Nervous Activity, A," 117. *See also* Artificial neural network; McCulloch, Warren; Pitts, Walter
Low-frequency analysis and recording (LOFAR), 197
Ludwig, Carl, 31

"Machine for Living" (Le Corbusier), 208
Machine learning, 5, 7, 79, 209, 251. *See also* Artificial intelligence
  in car sensing and driving technology, 140, 143
  definition of, 75
  to detect patterns, 184

  employed in Alter software, 108
  in e-tongues, 151
  and industrial food production, 157
  Kinect's use of, 73, 76–77
  in sleep trackers, 238
  in Spotify Running software, 166
Machine listening, 190
  relationship to Amazon Echo, 190–192
Machine perception, 136, 182, 185–186
  difference to human perception, 188–189
Magic lantern, 89
Magic Leap, 27–28
Maier, Tracy-Kai, 44
Mall of Dubai, 88
Maneuvering Characteristics Augmentation System (MCAS), 193. *See also* Boeing
Manpo-kei (10,000-step meter), 172
Marcuse, Herbert, 46
Marey, Étienne-Jules, 15, 18, 30, 33, 177
Marriage of Cadmus and Harmony, The, 43
Marshmallow Laser Feast, 87
Material imaginaries, 10
Mathematics, 4, 11, 39, 46, 183
  A-Life's use of, 109
  in analyzing sensor data, 176–177
  and automation, 38
  and calculus, 49
  in EEG analysis, 223
  Fechner's use to quantify senses, 13, 22, 254
  in machine automating of taste, 157, 159–160
  objectivity of, 9
  and psychology, 20
  use in Spotify's Running, 169
Mattel Power Glove, 66
Maturana, Humberto, 110
McCulloch, Warren, 116–117, 120–121
McCollum, Burton, 51

McLuhan, Marshall, 86–87, 105
McNamara, Robert, 198
McNamara's wall, 198
McQuaid, John, 151
Measurement, 7, 30–31, 39–40, 55–58, 137, 247–249. *See also* Psychophysics; Sensors
  benchmarking, 201
  and brainwaves, 220, 228, 236
  and calibration, 65
  Fechner's use of in psychophysics, 17–22, 26, 29
  and the Fun Palace, 104
  in industrial food industry, 157–158
  and information, 181
  instruments dedicated to, 18, 32–33, 35, 37, 51, 226
  physiological, 142, 166
  and the Quantified Self, 165–166
  of the self, 10, 101
  and sensation, 34
  through sensors, 25, 45–46, 49–50, 53–53, 57, 59, 67–68, 70, 73, 78, 85, 129–130, 140,143, 169, 171, 175–176, 181, 184–185, 192–193, 197, 204, 223 210–211, 235, 240–241
  in sensory science, 154–156
  sleep, 233–234, 239
"Medium Is the Message, The," 86
Meow Wolf, 87–88. *See also* House of No Return
Metropolis (film), 9
Microblindness, 27
Microelectromechanical systems (MEMS), 54–55, 72, 171
Microsoft, 28, 57, 139. *See also* Kinect
  invention of Kinect, 72–77
Millennium Bridge, 52
Mills College, 46–47
MindMachines.com, 238
Mist computing, 202, 204
MIT Media Lab, 32, 240
MIT Senseable City Lab, 208

Moholy-Nagy, László, 90–91
Moment Factory, 87
Monell Chemical Senses Center, 154
Monitoring, 7, 72, 104, 145, 158, 249
  self, 37–38
  with sensors, 173, 175, 180, 202–205, 209, 211–214, 234, 247, 251
  through sensors in cars, 131, 135, 139
  telemetry, 51
Mood enhancement, 136
Mood rings, 234
Mood sensing,139–141
Mood tracker, 6
Moodify, 140
Moore, Gordon, 54
Moore's Law, 54
Morphogenesis, 113–114
Mosso, Angelo, 35–36
Mouthfeel, 154
Mouth-smell, 152
Moxibustion, 16
Mud Muse, 97–98
Multimodal, 151, 170
  definition of, 203
Muscle atonia, 240
Musk, Elon, 130, 242
Mysterium, 91

Nafus, Dawn, 212
Nanna: The Soul Life of Plants, 20
Nanotechnology, 55
NASA, 68, 249
NASDAQ, 2
National Bureau of Standards, 50
Natural Born Cyborgs, 245
Natural interaction, 68–69, 74–75
  BMW's concept of, 140
Natural language processing (NLP), 191
Nature, 248
Nearables, 236
Neotame, 153
Netflix, 8, 115, 168

Network-centric warfare, 199
Neural interface, 233. *See also* Brain-computer interface (BCI)
Neuralink, 242
Neuromarketing, 242
Neuroplasticity, 112, 243
  definition of, 229–230, 232–233
  and sleep, 236
Neuron (biological), 30, 114, 153, 180, 186, 188, 220, 223–224, 227–229, 236, 238, 242–243, 251, 253–254. *See also* Artificial neural network; Spiking neural network
  artificial, 117–120
  components of, 116, 122–123
  and COVID-19, 250
  Hayek's description of, 118
  spiking, 121–121
  and taste, 180–181
Neuron (Arup software), 5–6, 209
Neuroscience, 26, 108–109, 222, 227, 232, 236
New Babylon, 196, 206, 208, 210–211
New York Times, 62, 135, 158, 174
New Yorker, 232
Next-Generation Nonsurgical Neurotechnology (research program), 242
Nieuwenhuys, Constant Anton, 195–196, 202, 206, 208, 210–211, 215
Nike, 176
Nintendo, 62–63, 66, 70–73
Nintendo Entertainment System (NES), 63
Nokia, 171, 173
Nolan, Christopher, 241
NTT (Nippon Telephone and Telegraph), 171–172
Nuremberg convention, 93

Oculus, 17, 25, 29
Ohm's law, 22
Office of Naval Research (ONR), 197

Office of the future (Xerox PARC), 68–69
OLED, 25
Onboarding (VR), 23
OpenBCI, 228
Open Score, 99. *See also* Rauschenberg, Robert
Operational closure (in biological systems), 110–111. *See also* Varela, Francisco
Orthosomnia, 238
Orwell, George, 103
Oscillators, 122
Oscillopsia, 26
Outside in sensors, 25. *See also* Sensors

Panpsychism, 19
Parameters (sound), 3, 6, 25, 48
  in mathematics, 160
Park, James (Fitbit founder), 177
Park Avenue Armory, 99–100
Pask, Gordon, 101–103. *See also* Fun Palace
Patents, 12
  accelerometry, 175–176
  BMW face recognition, 140
  Spotify, 166
Path dependency, 71–72, 119
Pattern, 6–7, 39, 96, 112, 120, 173, 182–185, 187, 189, 191, 194, 204, 238, 244, 248. *See also* Pattern recognition; Turing patterns
  brain machines generating of, 238
  classification of, 185, 188
  in data, 7
  definition of, 182
  in lucid dreams, 240
  in sleep, 6
  in neuronal spikes, 117–118, 121
  and self-organization, 123
  in TVSS, 230–231
  use of in Kinect infrared light, 76

Index 313

Pattern recognition, 13, 148, 182–183
Peaks (in data), 184
Pedometer (step counter), 171
   deployed in mobile phones, 173–174
   origin of, 172
   use in Japan, 172–173
Pepper's ghost technique, 89
Percept, 203
Perception-action cycle, 130, 132, 134,136,144
Perceptron, 255. See also Rosenblatt, Frank
Persephone myth (*Eidos:Telos* ballet), 44–46
Pervasive computing, 202
Peters, Orville, 51
Phantasmagoria, 89
Phenomena, The, 87
Photographic gun (Marey), 29. See also Chronophotography
Physics, 13, 46–47, 50
   Fechner's study of, 15, 19
   and nanotechnology, 54–55
   relationship to accelerometer, 50, 78–79, 157, 242
Physiology, 15, 143, 148, 214, 240, 248
   neuro-, 102
   sensory-, 30, 32, 34
Pine, Joseph, 87
Pitts, Walter, 116–117, 120–121. See also McCulloch, Warren
Placebo effect, 243
Plethysmograph, 35–36, 225
Polhemus, 67–68
Polysomnography, 235
Popova, Lyubov, 91
Porsche, 138–139
Poses, 73
Power Gig (game controller), 62
Prediction, 8, 105, 128, 131, 182, 205, 213–214, 251, 278n2
   in pattern recognition, 185–186

Predictive analytics, 7
Preventative health, 213
Price, Cedric, 101–103
Princeton/Columbia Electronic Music Studio, 47
Processed food industry, 152. *See also* Sensory science
   and mouthfeel research, 154
   and sensor technology, 157–160
   use of artificial flavors, 153
   use of sensory science, 154–157
Project Nine, 198. *See also* Igloo White
Psychoacoustics, 137
Psychology Review, 255
Psychology Today, 219
Psychophysics, 17, 100, 226
   definition of, 18–19
   Fechner's invention of, 22, 39–40
   and method of adjustment, 23
   and method of limits, 23
   and method of right/wrong cases, 23
   and psychoacoustics, 137
   relationship to VR testing, 25–29
   use in sensory science, 156
   in Wundt's lab, 30, 32
Puja, 9
Pulse oximeter, 1. *See also* Sensors
   definition of, 271n30
   developed by Nintendo as video game controller, 79
   in Dreem 2 EEG headset, 236
   exhibiting racial bias, 1
Punchdrunk, 87

Quantified self, 174–175, 177–178, 245

Raku-Raku Fujitsu phone, 172–173
RAND Corporation, 98, 234
Rapid eye movement (REM), 27, 240
Ratti, Carlo, 208

Rauschenberg, Robert, 96. *See also Mud Muse*; *Soundings*
    collaboration with Teledyne, 97–98
    founding of E.A.T., 100
    and *Nine Evenings*, 99–100
    and reactive environments, 96–97
Reaction-diffusion, 112–114
Reaction time, 31, 33–34, 40
Receptor cells, 148, 152
Recommender systems, 168
Reddit, 1, 166
Redirected walking, 27
Reference standard, 65
Regression (statistical process), 159
Remote sensing, 203
Research and Development Center for Five-Sense Devices (Kyushu University), 147
Resonance, 52, 168–169
Response time (in psychology), 18, 184. *See also* Reaction time
Re-Timer, 237
Retronasal smelling, 152. *See also* Mouth-smell
Rheology, 157
Riefenstahl, Leni, 92–93, 100
Rock Band 3, 62
Roomba (robotic vacuum), 4
Rosenblatt, Frank, 255. *See also* Artificial neural network; Hayek, Friedrich; Perceptron
Running (Spotify app), 166–169, 173, 177
Ryan, Joel, 45, 55
    definition of real time, 47
    use of accelerometer, 48–50, 53

Saccades, 25, 27
Sample resolution, 184
Sampling rate, 114, 205
Samsung, 24, 28
San Francisco, 230
    heating up during COVID-19, 249
    and high-tech monopolies, 9
    use of wireless parking sensors, 135
Scary Beauty (robot opera), 108, 123
Science, 205
Security, 6
Sega, 64–66
    Activator (game controller), 64–66
    Genesis, 64
Selectivity, 148
    comparison to global and high selectivity, 148
Self-governance, 110
Self-organization, 109, 119
    as principle of artificial life, 123
Self-registering (instruments), 36
Sensed self, 10, 13
Sense (sleep tracking company), 236
Sensation, 9, 12, 29, 31–32, 214, 250
    definition of, 18
    Fechner's understanding of, 20–23
    Hayek's theorization of, 118–119, 253–254
    in machines, 182, 188–189
    modeling of, 26
    and taste, 153–156
    Wundt's study, 33–35
Sense making, 109, 119, 123–124, 253
Sensing machines, 9, 11, 29, 36–37, 39–40, 43, 75, 78, 90, 102, 104–106, 110, 179, 189, 195, 225, 251–252, 254
    Amazon Echo as, 192
    cars as, 135–136
    in contemporary psychophysics research, 19, 26
    definition of, 4
    enabling immersion, 87
    fitness trackers as, 174, 177–178
    and human desire, 10
    imagined by twentieth-century artists, 90, 94–96, 101–102
    and infrastructures, 5
    as musical instruments, 56, 58

Index

in the nineteenth century, 18, 30–31, 177
relationship to environment, 6
role in COVID-19 pandemic, 14, 248
shaping the self with, 223, 233, 243, 245–246, 254–255
and sleep, 234, 236
and taste, 151, 157, 159–161
video games as, 63, 72, 79
in Wilhelm Wundt's lab, 32
wireless sensor networks as, 202–203
SensIT, 199
Sensor field, 201
Sensor fusion, 129
definition of, 130, 278n3
Sensorimotor, 68, 75, 111, 180–182
Sensor Lab (control system from STEIM), 45, 50
Sensors, 1. *See also* Accelerometer; Pulse oximeter
angle of attack, 192
bathythermograph, 197
breathalyzer, 250\
carbon dioxide (CO_2)
definition of, 5
electromyogram (EMG), 142
electrodermal activity (EDA), 142
flip flop motion detector, 170
galvanic skin response (GSR), 142
gyroscope, 5, 135, 165, 168, 225, 248
inertial measurement unit (IMU), 171
infrared (IR), 6, 64, 73, 76, 134, 240
infrared thermometer, 157
motion, 78
nanoscale, 150
near-infrared microspectrometry, 161
piezoelectric, 5
photocell (photoelectric or Light Resistive Diodes), 58, 67, 95, 99
seismic intrusion detectors (SID), 199
sound ambient level sensor (SALS), 137
spectrometers used in food, 161
tilt, 5, 73, 130, 171
time constant of, 184
ultrasound, 70
Sensors (journal), 249
Sensory science, 154–157
Sensory Order, The, 117, 120, 253–255, 276n22. *See also* Hayek, Friedrich
relationship to economics, 277n28
Sensory substitution, 230
military interest in, 232
origin of, 230
popularization of, 232
Shankar, Ravi, 46
Shannon, Claude, 54
Sharon, Yigal, 140
Shelley, Mary, 9
Shepherd, Gordon, 152
Shibuya, Keiichiro, 108
Sidewalk Labs, 208, 210
SIGMA+, 199
Sine wave, 48
Singapore, 1
as sensor city, 210–211
teamLab installation in, 89
Situationist International, 195
Skrjabin, Aleksandr Nikolajevič, 91
Sleep, 233
health effects from lack of, 233–234
and product claims, 234
purpose of, 236
and wearable tracking devices 234, 236–238
Sleep apnea, 235
Sleep machines, 6, 238
Smart car sensor network, 135
Smart Dust, 202
Smart-Its, 202
SmartMesh, 202
Smith-Kettlewell Eye Research Institute, 230
Soft self, 246
Solenoid, 230
Songs and Stories from Moby Dick, 57

Sony, 72, 138–139
  development of Walkman and Watchman, 168
Soundings, 96, 100
Sound source localization, 190
Sound Surveillance System (SOSUS), 197–198
Sound symposer, 138
Spackman, Christy, 157
Speech recognition, 99, 182, 191, 250
Sphygmograph, 38
Spike buoys, 198
Spiking neural network, 112, 121
  Izhikevich model, 121
Spotify, 115, 143. *See also* Patents; Running
  Running app, 165–169, 173, 178
Stanford University, 143, 240
Statistics, 38, 157, 178, 184
  origin of, 7–8
Sterling, Bruce, 169
Stimuli, 6, 11, 20, 29, 35, 37, 114, 133, 141, 158, 160, 233, 240, 253
  definition of, 18
  Fechner's distinguishing from Sensation, 20, 26, 30
  Hayek's theory in relation to classification, 118, 120–121, 253–255
  machine sensing and, 185, 188, 190,
  method of constant, 23
  organisms sensing of, 180, 183
  and shock (Benjamin), 92
Stoics, 175–176
Strange Days (film), 241
Stravinsky, Igor, 45
Stress, 141
  definition of, 141
  Just in Time reduction, 144–145
  measuring with sensors, 141–142
  physiological features, 142
Structural coupling, 109
  definition of, 111

enacting a world through, 124
  relation to sense making, 123
Structured light, 76
STEIM, 45–46, 56, 267n4
Superblue, 87
Surveillance, 6
  Fun Palace's version of, 103
  in military use, 197, 199, 201
  with neurotechnologies, 243
  and sensors, 7, 199, 211
Surveillance capitalism, 9, 251
Symbiosis, 252
"Systems Esthetics" (Burnham), 95

Tachistoscope, 18, 35
Tactile Vision Substitution System (TVSS), 230–232
  operation of, 231
Taste, 148. *See also* E-tongue
  e-tongues' replacement of, 149, 151
  and flavor, 151
  physiology of, 148, 152
  relationship to smell, 152
  sensors, 147–148
  and tastebuds, 148
Taste modulators, 7, 154
teamLab, 83–87, 105–106
Teledyne, 97–98
Telemetry, 50
Tesla, Nikolai, 52
  earthquake machine, 52
Theater of the mechanical eccentric, 90
Theater of totality, 90
THIM (sleep tracker), 237
This, Hervé, 160
Tickets (restaurant in Barcelona), 150
Time sense apparatus, 34
Time series, 184–185
Titchener, E. B., 35
Toko, Kiyoshi, 147–150
Topology, 201

Index

Tracking, 13
  motion, 76, 78, 85
  self-tracking, 174–175
  by way of sensors, 25, 27, 67–68, 73, 139, 176, 197–199, 202, 212, 237–238, 240, 248, 249
Training set (in Neural Networks), 75
Transcranial magnetic stimulation (TMS), 227–229, 233
Transducer, 158
Triangulation, 70
Trigeminal nerve, 152
Triumph of the Will, 92–93
Trumbull, Douglas, 241
Two-point threshold, 21
Tuchman, Maurice, 97
Tufte, Edward, 32
Turing machine, 111, 113
Turing patterns, 113–114
Turrell, James, 100–101

Ubiquitous computing, 69, 140, 208
Umwelt, 188
US Department of Defense (DOD), 198–199
US Department of Transportation National Highway Traffic Safety Administration, 141
Utzon, Jørn, 209

VaMorS, 127. *See also* Dickmanns, Ernst
Varela, Francisco, 110–111
Varian, Hal, 8
Variations VII, 99–100. *See also* Cage, John
Vector, 51
Venice Biennale, 95
VJ (video jockey), 64–65
Vietnam War, 198
  early use of sensor networks, 198–199
  origin of electronic warfare, 198

Virtual reality, 13, 17, 23, 69
  Artaud's definition of, 67
  as real virtuality, 28
  use of psychophysics in, 25–29
  and the VPL Data Glove, 67–70
Voice sniffing, 191
VPL (Silicon Valley Company), 67–68. *See also* Data glove (VPL)

Waldhauer, Fred, 100
Walker, Matthew, 235
Walmart, 234
Wakeword (used for Amazon Echo), 190–191, 288n25
Weber, Ernst Heinrich, 21, 23
Weber's law, 23
Weiser, Mark, 68–69. *See also* Ubiquitous computing
Western Electric, 196
Westmoreland, William, 199
Whitman, Robert, 99–100
WHO (World Health Organization), 132, 161, 233
Wiimote (or Wii Remote), 62, 64, 71–73, 77–79
Wiener, Norbert, 102, 189
Wiesel, Torsten, 186
Willems, Thom, 45
Window (signal processing), 38, 185–186, 235, 174
Wireless local area network (WLAN), 211
Wireless sensor network, 201–202, 211
Wolf, Gary, 174, 177
  founding of quantified self, 174
Wortz, Edward, 100
Wundt, Wilhelm, 32–35
  and introspection, 34
  invention of physiological psychology, 32
  psychological institute laboratory, 33

Xbox 360, 73, 75
Xerox PARC, 68, 242

You Are Not Your Brain, 244

Zapper light gun, 63
Zend-Avesta: On Matters of Heaven and the World Beyond, 20
Zeta music company, 57
Zimmerman, Thomas, 67